Marek Sokolski
Editor

Mining Machines and Earth-Moving Equipment

Problems of Design, Research and Maintenance

Springer

Editor
Marek Sokolski
Wrocław University of Science
and Technology
Wrocław, Poland

ISBN 978-3-030-25480-3 ISBN 978-3-030-25478-0 (eBook)
https://doi.org/10.1007/978-3-030-25478-0

This Springer imprint is published by the registered company Springer Nature Switzerland AG
The registered company address is: Gewerbestrasse 11, 6330 Cham, Switzerland

This book is dedicated to the memories of
Prof. Henryk Hawrylak and Prof. Kazimierz
Pieczonka
our Masters

Preface

In the 1960s, at the Department of Mechanical Engineering of the Wrocław University of Science and Technology (WUST), two new areas of research and education were established, “Mining Machinery” and “Earthmoving Equipment”. Two pioneers of these new research directions were Profs. Henryk Hawrylak and Kazimierz Pieczonka, both closely connected to the University since the University was established in 1945.

This book was compiled by their former students and co-workers to express the deep gratitude and fond memories about these two professors. The chapters in this book contain the examples of research created under either their direct supervision or initiated by them by offering new ideas.

Professor Henryk Hawrylak, born in 1925, came to Wrocław in 1945, just after the end of the Second World War and started studying at the Department of Mechanical and Electrical Engineering of the Wrocław University of Science and Technology. During his student years, he was very active, serving the academic community in the student organizations “Academic Guard” and “Brotherhood for Help”, both of them helping the university community to cope with difficulties of everyday life in post-war Wrocław.

He obtained his Ph.D. degree in 1961 and his D.Sc. degree in 1965. His D.Sc. dissertation on “The theoretical basis of design and selection of the working system of bucket wheel excavators” catalyzed new avenues for research and development of the entire branch of the heavy machine design and established the scientific school which Prof. Hawrylak created and managed for the subsequent 25 years. He was nominated the Associate Professor in 1969 and the Full Professor in 1975.

Professor Hawrylak significantly contributed to the development of science and engineering both at the university and in Poland. During his tenure at the WUST, he served in many administrative positions. Among them, he was the Vice-Dean and then Dean of the Department of Mechanical Engineering and the Vice-President (Scientific Staff Development). During the period 1988–2006, he was a member and subsequently the vice-chair of the Polish Central Commission for the Scientific Degrees and Scientific Titles. He was the theses supervisor of 16 Ph.D. candidates.

Prof. Henryk Hawrylak (on the left) and Prof. Kazimierz Pieczonka (on the right)

Out of his students, there are currently five full professors and four scientists with the D.Sc. title. He was the author or co-author of more than 110 scientific papers, eight books and textbooks and more than 100 major reports for the Polish industry.

In the period between 2000 and 2005, three Polish Universities, Wrocław University of Science and Technology, Koszalin University of Technology and Lublin University of Technology, bestowed the highest scientific honour of Doctor Honoris Causa on Prof. Hawrylak in recognition of his contribution to science, technology and industry.

In additional to his academic activities, Prof. Hawrylak was very closely connected to the Polish open mining industry. He was a chairman of the Advisory Board for Design Office Poltegor. In this capacity, he significantly affected the development of the Polish open mining industry contributing considerably to the Polish economy and making sure that the Polish energy sector was self-sufficient. In recognition of his contribution, he received a title of an honorary employee of all Polish open-pit brown coal mines.

Professor Hawrylak passed away on 21 November 2013.

Professor Kazimierz Pieczonka, born in 1925, came to Wrocław in 1948 and started studying at the Mechanical Engineering Department of Wrocław University of Science and Technology. He graduated in 1954. During his study programme, he was employed as the teaching assistant and the research assistant at the Cathedral of Machine Design of the Department of Mechanical Engineering.

He defended his Ph.D. theses in 1964 and in 1986 became an Associate Professor at the WUST. He became a full professor in 1993.

During his tenure at the Department of Mechanical Engineering of Wrocław University of Science and Technology, Professor Pieczonka served the university community in many important roles. Among others, he managed the Teaching Group of Heavy-Duty Machinery, he was Scientific Editor of the Transactions of the Institute of Machine Design and Operation, as well as the Chairman of the Permanent Committee for Doctoral Studies.

His scientific research revolved around bulk solids handling, loading, unloading and moving of granular media and also terramechanics of the various types of earthmoving machines on tyres.

From the very beginning of his long scientific carrier, he was closely connected to the Polish industry, particularly those companies that were located in Lower Silesia. Particularly, he was engaged in close collaboration with Fadroma, Wrocław and KGHM, Lubin, as well as with the Wrocław Branch of the Industrial Institute of Mechanized Construction and Rock Mining.

Professor Pieczonka created the school of theory of design and operation of self-propelled earthmoving machinery in Wrocław. He was one of the best *authorities* on this subject.

His achievements were documented in more than 100 scientific papers, two books, two textbooks and more than 100 industrial technical reports. He supervised six Ph.D. theses. Two of his students became full professors, in addition to one with the title of D.Sc.

Professor Pieczonka passed away on 3 August 2017.

Wrocław, Poland Marek Sokolski

Contents

Resilience Engineering—Agents of Open Pit Mining Machine Disasters in Poland

Dionizy Dudek and Tomasz Nowakowski

Abstract Resilience is defined as readiness for secure and acceptable service under abnormal work conditions (e.g. disruptions, attacks, accidences, disaster). The paper shows possibilities of disruption process analysis for real engineering machines. The expert analysis of disasters creates a unique opportunity to address the real catastrophic events concerned excavators and spreaders in Polish open pit mining. Total number of undesired events has reached 235 and affected 95 machines. Only 23% of all machines were not involved in a large breakdown. The most frequent catastrophic events dealt with impact load (120 events) and failure of the system rotating machine body (67 events). The problem is described by 6 particular cases.

Keywords Resilience · Machine · Disaster · Open pit mine

1 Introduction

The issue of design process [1] is traditionally placed on physical protection of the object and asset hardening [2]. However, in recent years, lessons learned from some catastrophic accidents have pushed part of the focus on the concept of 'resilience' [2]. The results of different accident investigations pay attention to the need for introducing the term 'resilience' into the disaster discourse, giving birth to a new culture of disaster response [3]. Consequently, government policy has also evolved to encourage efforts that would allow assets to continue operating at some level or quickly return to full operation after the occurrence of disruptive events [4].

'Resilience' comes from the Latin word 'resilio' that literally means 'to leap back' and denotes a system attribute characterized by the ability to recover from challenges or disruptive events [5]. The Merriam-Webster dictionary defines resilience as 'the ability to recover from or adjust easily to misfortune or change'. In this view, systems should not only be reliable, i.e. having an acceptably low failure probability, but also

D. Dudek (✉) · T. Nowakowski
Faculty of Mechanical Engineering, Wroclaw University of Science and Technology, Wybrzeze Wyspianskiego 27 (ul.), 50-370 Wroclaw, Poland
e-mail: dionizy.dudek@pwr.edu.pl

M. Sokolski (ed.), *Mining Machines and Earth-Moving Equipment*,
https://doi.org/10.1007/978-3-030-25478-0_1

resilient, i.e. having the ability to optimally recover from disruptions of the nominal operating conditions [2, 5].

As an example of the analysis of the concept of resilience, one can show the review of transportation process or logistics/supply chain process resilience assessment methods [6]—Fig. 1.

To sum up the comprehensive literature review [6], there can be presented the three main definitions of conceptions connected with the effective performance of supply chains which are vulnerable to disruptions.

Reliability assessment is focused on the possibility of an unwanted event occurrence. Measures of the reliability should express uncertainty about the appearance of such an event, like failure, fault, error, etc. Thus, reliability (dependability) of a logistic system can be understood [7] as the ability to deliver correct service under normal (ordinary) work conditions in a given time interval.

Safety means the absence of critical/dangerous events while security is focused on protecting the system environment against the effects of these damages. Safety is measured generally by risk—two-dimensional combination of probability of an undesirable event and possibility of loss (consequences). Risk assessment consists of process of risk identification related to threat, includes its possibility (likelihood or probability), impact and consequences.

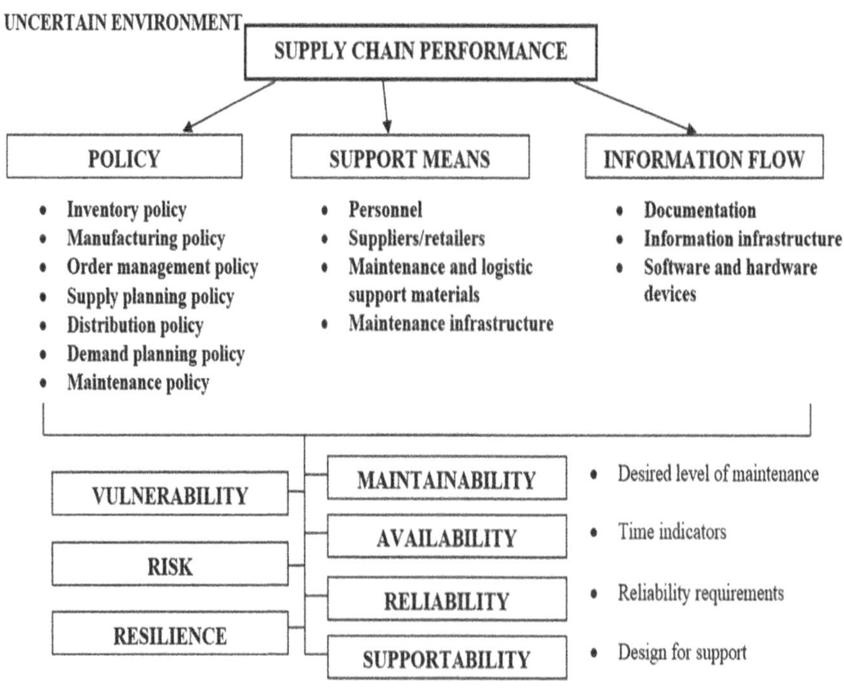

Fig. 1 The key characteristics in achieving supply chain performance [6]

Resilience takes into consideration not only the discussed issues (reliability and safety) but also the possibilities of restoring the original properties of the system. Thus, resilience means readiness for secure and acceptable service under abnormal (uncommon) work conditions (e.g. disruptions, attacks, accidences, disaster). The measure of resilience can be understood as time to restore the capabilities of the system (worse than new, as good as new, better than new [7]).

2 Disruption Process

In the current literature, there can be found research works dedicated to vulnerability and resilience measurement issues. The vulnerability assessment issues are investigated in detail by the authors in [8]. The examples of supply chain resilience measurement systems/methods are given in [9], where authors present two resilience-based component importance measures for networks, in [10], where authors investigate the assessment issues of passenger transportation system's resilience, in [11], where a method for measuring resilience based on fuzzy logic is proposed, and in [12], where authors develop generic metrics for quantifying system resilience. In work [13], authors introduce a resilience metric that incorporates the three resilience capabilities (absorptive capacity, adaptive capacity, recovery capacity) and the time to recovery. Following this, in Fig. 2, there is presented the resilience measurement framework.

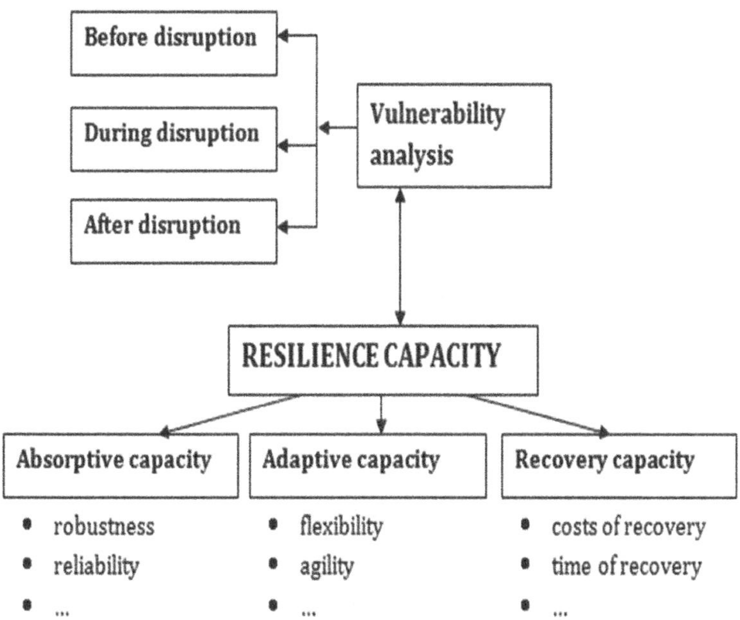

Fig. 2 Resilience framework [6]

Moreover, the complexity of the problem is connected with the necessity of taking into account some factors that might increase the level of risk, like [14]:

- focusing on efficiency targets instead of effectiveness issues,
- supply chain globalization,
- focusing on factories and centralized distribution,
- outsourcing,
- reduction of the supplier base,
- demand variability and
- lack of visibility and control procedures.

These factors are discussed in more depth, e.g. in [15].

Any significant disruption will have a typical profile in terms of its effect on company performance. The performance is measured by sales, production level, profits, customer service or another relevant metric [8].

The nature of the disruption and the dynamics of the system response can be characterized by the following eight phases (Fig. 3) [16]:

1. Preparation. In some cases, a company can foresee and prepare for disruption, minimizing its effects. In other cases, there is little or no warning.
2. The Disruptive Event. The tornado hits, the bomb explodes, a supplier goes out of business or the union begins a wildcat strike.
3. First Response. Whether there's a physical disruption, a job action or an information technology disruption, first response is aimed at controlling the situation,

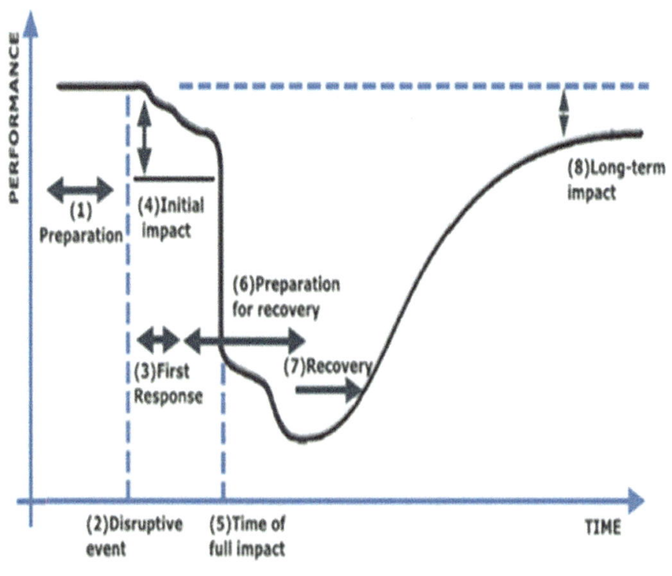

Fig. 3 The disruption profile [17]

saving or protecting lives, shutting down affected systems and preventing further damage.

4. Initial Impact. The full impact of some disruptions is felt immediately. Other disruptions can take time to affect a company, depending on factors such as the magnitude of the disruption, the available redundancy, and the inherent resilience of the organization and its supply chain. During the time between the disruptive event and the full impact, performance usually starts to deteriorate.
5. Full Impact. Whether immediate or delayed, once the full impact hits, performance often drops precipitously.
6. Recovery Preparations. Preparations for recovery typically start in parallel with the first response and sometimes even prior to the disruption, if it has been anticipated. They involve qualifying other suppliers and redirecting suppliers'.
7. Recovery. To get back to normal operations levels, many companies make up for lost production by running at higher-than-normal utilization, using overtime as well as suppliers' and customers' resources.
8. Long-Term Impact. It typically takes time to recover from disruptions, but if customer relationships are damaged, the impact can be especially long-lasting and difficult to recover from.

Chapter "Modelling, Computing, and Analyzing Large-Size Rotary Joints" shows the possibilities of disruption process analysis for real engineering machines. Is it practicable to identify all the phases shown in Fig. 3? More detailed statistical analysis of the problem is very limited by the lack of certain information. Often, the data are regarded as private or secret and results of analysis are impossible to be published [7]. So the expert analysis of disasters discussed in [1] creates a unique opportunity to address the real catastrophic events.

3 Engineering Machines Disasters

In Polish open pit mining, there were and exist now 124 machines of 46 different types including excavators and spreaders. After World War II, it was observed 235 breakdowns and 38 of which were classified as major catastrophes related to 95 machines. Undesired events had both design and operational causes. A total number of multiple failures caused by five structural units reach 205. Another 30 failures had operational and environmental causes. Each of 235 catastrophes was followed by official penetrating inquiry looking for basic causes [18].

It was observed 87 different events classified as minor (59) and major (38) catastrophes. Some of the failures repeated more than once and happened even in the same machine. Total number of undesired events has reached 235 and affected 95 machines. Only 23% of all machines were not involved in large breakdown [1, 18].

The most frequent catastrophic events dealt with impact load (120 events) and failure of the system rotating machine body (67 events). More precise analysis shows just for these events the following failure modes:

- impact load:
  - break of the bucket wheel boom structure,
  - deformation of the bucket wheel boom structure,
  - break of the beam supporting bucket wheel gearbox,
  - break of gearbox cogs,
  - crack of the wheel boom structure and superstructure masts,
  - crack of the gearbox and bucket wheel shaft,
- system of the machine body rotating (solid-web girders: lower at undercarriage and upper supporting superstructure),
  - crack of undercarriage supporting portal,
  - crack of welds and structure of solid-web girders.

Failures caused by excessive filling of the transfer hopper with the excavated material being transported, which is commonly referred to as 'excavated material overfilling', should officially not occur. Design standards for basic machinery recommend that an additional load due to the deposited excavated material should be included in the strength calculations of load-bearing structures, too. This relates to the material deposited:

- in clogged chutes of the bucket wheel and transfer hoppers, with the excess material resulting from the natural angle of the excavated material heap;
- on conveyor belts with side barriers, up to the top height of such barriers;
- on the load-bearing structure of conveyors, extension arms, the chain or the bucket wheel in the form of clinging material or soiling;
- on platforms in the vicinity of the clogged transfer hopper.

However, the examples discussed in the following text conclude that standardized recommendations do not necessarily go hand in hand with the design concept and the daily routine in exploitation.

It must be emphasized here that abrasive processes at the cladding/transported material boundary considerably contribute to this type of failure [1]. This relates to the type of cladding material used for chutes or the hopper, i.e. its abrasion resistance and impact strength as well as the tendency to become covered with clinging material. Mining rocks of various workability classes within individual geological strata necessitates such technical solutions that can guarantee that the maximum resistance to abrasion and impact strength would be obtained. A vast percentage in the general mass of the mined material at lignite mines consist of soils (difficult in terms of workability), which considerably exert abrasion impacts on the cutting elements of buckets as well as cladding elements in the chutes and transfer hoppers.

The intensity of this process depends on the content of hard particles from materials which form the soil structure, in particular, quartz with sharp-edged particles (firmly deposited in the structure of the rock being mined). As a result, the cladding is excessively worn, hence geometric shapes change and conditions for collecting the excavated material stream become worse. In addition, the machine operator is not blameless when overfilling occurs. It is mainly formed when [1]:

- the technical condition of the chutes in the transfer hoppers is of poor quality, which is not that rare in mines;
- instantaneous output of the machine considerably exceeds the rated value, often the theoretical yield as well, which is almost a daily occurrence in the case of light sandy loam;
- extraordinarily unfavourable weather conditions exist;
- the later examples of causes originate from post-failure machinery documentation after heavy—even pouring—rain.

Among other subassemblies of open pit mining machinery especially endangered to wear caused by abrasive processes are undercarriages. In friction nodes under high values of loading hard soil particles can initiate micro-cracks leading to chapping of surface layer of undercarriages' elements. Such degradation can run gradually or in some cases at an accelerated rate substantially limiting the durability of those parts. This problem is discussed in details in [19].

3.1 Stacking Machine A2RsB-5000—Case I

A 14-year-old stacking machine A2RsB-5000 with the theoretical yield of $Q_{theor} = 5000\ m^3/h$ and a weight of $G = 21.60$ MN was operated at the end of August during the morning shift while stacking the excavated material during extremely torrential rain. The excessively watered material load on the entire belt conveyor—lifted at an angle of approx. 16° by wet belt no. III of the stacking machine—lost its adherence to the belt and started to slide down towards the transfer hopper of belt no. II on the loading elevator. In that situation, stacking machine belt no. III stopped collecting the excavated material from transfer point II/III. This resulted in the extreme filling of the transfer hopper; consequently, the load-bearing structure of the loading machine fixed jib broke (Fig. 4).

The break happened due to the too low strength of the poles relative to this extreme weight. The posts twisted around the axis by approx. 80°, and simultaneously the lower belts of the jib, fastened frontally to the posts as rigid joints, were torn. This can be clearly observed in Fig. 5. The break occurred along an approx. 15.5 m long section of the load-bearing structure of the jib, which additionally bent by approx. 5 m. The following also experienced some considerable damage:

- upper and lower belts as well as posts and cross-beams;
- load-bearing structure of the belt no. II conveyor;
- structure of the roller supports of belt no. III;
- longitudinal circulation platforms on the fixed jib;
- transverse circulation platform on the fixed jib.

Fig. 4 Stacking machine A2RsB-5000—Case I. Broken machine due to material slippage and the resulting overfilling—loading machine fixed jib. The arrow indicates the failure spot

Fig. 5 Broken fixed jib due to material slippage and the resulting overfilling

During the 35-day repair, the load-bearing structures of all jib elements as well as platforms were replaced.

Searching for this particular case of loading in standard regulations failed. Consulting this issue with the stocking machine manufacturer and strength calculations also concluded that no case of extreme loading like the mentioned one had been envisaged. However, the problem continues to exist, as violent downpours in Poland during the spring and summer periods occur, and a few stacking machines of this type are still being operated.

3.2 Stacking Machine A2RsB-5000—Case II

What goes around always comes around. This time, a 26-year-old stacking machine of the same type was being operated at the beginning of July to stack up materials. At the end of the second shift, violent rainfall, literally a downpour, took place. Also at this time, the over-wet (almost fluid) excessive load on the entire belt conveyor slid down to the return drum once the load lost adherence to belt no. III, and it fell (leaked) onto the working floor (Figs. 6 and 7).

In this situation, belt no. III completely stopped collecting material from the hopper at the transfer of belts II/III. Thus, it immediately was filled up with the material being conveyed by belt no. II, and the so-called excess covering occurred as well (Fig. 8). As a result, the load-bearing structure of the fixed jib, supporting

Fig. 6 Stacking machine A2RsB-5000—Case II. Right-side view on the broken fixed jib

Fig. 7 Stacking machine A2RsB-5000—Case II. Left-side view on the broken fixed jib

belt no. II (loaded with an excessive amount of wet material being transported) was broken by the additional weight, similarly to the previously-described case, which took place 11 years earlier.

This time the amount of the semi-fluid material, which accumulated at transfer point II/III, was considerably larger, which had an obvious impact on a wider scope of damage, which was experienced by the following:

- broken protection against the rotation of belt no. II on the conveyor at its suspension point;
- bent upper belts of belt no. II conveyor structure and bent jib beam;
- damaged support carriage of the structure of belt no. II on the conveyor;
- bent elements of the fixed jib structure;
- support posts twisted around the axis by approx. 80°;
- lower belts of the fixed jib torn away from the frontal fastening to the posts;
- supporting posts twisted similarly as in Case I;
- similarly, rigid joints connecting the posts with the lower belts of the loading elevator bodywork were torn up;
- fixed jib structure—once the support was lost—collapsed at a length of 18 m and bent by approx. 7 m, until it rested on the working platform of the stacking machine with rectangular frame connecting it with the belt II/III transfer point.

Fig. 8 Stacking machine A2RsB-5000—Case II. The return drum and the frame supporting belt no. II structure at the level of the stacking machine. View of the wet material after flowing down from the belt

The excessive bending of the structure of belt no. II on the conveyor also caused breakage of the protection against conveyor rotation.

This was another specific, almost identical, case of loading the stacking machine transfer points by wet material flowing down the belts. The reinforcement of the posts implemented after the first failure (11 years ago) had no effect. It is unsatisfactory that strength calculations do not envisage these types of loads. Such failures have already reoccurred three times. These are very troublesome and expensive situations. Also, the repair may take a long time. It appears that the compiled standards should be extended by such a special loading case.

3.3 Stacking Machine A2RsB-5000—Case III

This time, a 31-year-old stacking machine A2RsB-5000 suffered a similar failure under very comparable circumstances. Again, in the summer of mid-July during the night shift, when it was heavily raining, the excessive load slipped and slid into the hopper at the transfer point between belts II/III. The angle of inclination of belt no. III was approx. 18°. Similar to the two previous cases, the excessive load accumulated in the filled hopper broke the load-bearing structure of the loading elevator fixed jib. Although it is true that the reinforcement of the fixed jib structure implemented within a period of six years turned out to be the right solution (95%), both lower belts of the structure within cross-sections failed. They experienced distinct buckling and then fracture as well. Additionally, the rigid joints connecting frontally with the posts broke off. The previous reinforcement of the posts within the cross-section proved to be the right solution, and the insufficiently high strength of the posts contributed to this considerable damage in the previous cases.

Fig. 9 Stacking machine A2RsB-5000—Case III. Right-side view on the broken fixed jib

Fig. 10 Stacking machine A2RsB-5000—Case III. Broken loader elevator fixed jib viewed from the stacking machine side

The loading elevator fixed jib without support in the area of the lower belts broke the upper belts at a length of approx. 3 m and fell onto one side by approx. 6.5 m until it rested on the floor by means of the rectangular frame. In addition, the damage was experienced by:

- circulation platforms;
- the transfer hopper associated with belts II/III;
- the travelling path of the bogie of belt no. III;
- main belts of the structure of the extension arm associated with belt no. III;
- load-bearing structure of the conveyor with belt no. II;
- slide rings in electrical equipment in the upper part of the stacking machine (Figs. 9 and 10);

The repair of the damaged machine lasted 49 days and consisted of the following:

- repairing all the listed damage;

- implementing a cross-beam and four slanting diagonal stays in cross-sections, thus considerably reinforcing the lower belts;
- implementing two additional triangle joint metal sheets to stiffen the connections of the belts with the poles.

Doesn't the repeated occurrence of these three failures seem odd? Fortunately, for the currently designed stacking machines, design engineers have stopped to apply a solution, wherein the extension arm of belt no. III is suspended through the rectangular frame to the fixed jib of the loading elevator. The analyses of repeatable failures, but not limited to, has proven to be good training for both design engineers and operators.

3.4 *Stacking Machine A2RsB-5000—Damaged Joint of Stacking Machines with Loading Elevators*

A 25-year-old, two-member stacking machine (with a loading elevator) was operated to stack horizontal heaps (Fig. 11). In the first half of May, when the weather was very good, both machines were ascending along the feeding conveyor. The inclination of the working level was as follows: at the loader elevator—1:15.8, at the stacking machine 1:14—for the value allowable by standards and conforming to the operating instructions for both machines—1:30. When they stopped, the brakes on the caterpillar vehicles of the loading elevator failed to stop along this inclination, and uncontrolled back-movement of the machine by approx. 15 cm resulted.

Fig. 11 View on the stacking machine-loading elevator assembly (A2Rs.B-5000) before the failure

The load-bearing structure of the conveyor with belt no. III moved outside the supporting trolley on the loading elevator and fell down onto the working level from a height of approx. 8 m (Fig. 12). This incident caused the following:

- broken upper and lower belts and struts of the conveyor load-bearing structure at a length of approx. 10 m;
- bending of the final part of the conveyor load-bearing structure associated with belt no. III;
- ruptured a bunch of rubber-coated power and control cables between the stacking machine and the loading elevator;
- bending of vertical posts associated with suspension of the conveyor structure on the loading elevator;
- damaged circulation platforms along the conveyor structure;
- bending of the transfer hopper from belt no. III to belt no. IV;
- damaged conveyor structure associated with the cleaning belt under belt no. III.

The damage repair took 16 days.

Fig. 12 View on the stacking machine-loading elevator assembly (A2RsB-5000) after the failure

3.5 Stacking Machine as-50.12—Bodywork Tilting

A 23-year-old stacking machine with the theoretical yield of $Q_{theor} = 3080\ m^3/h$ and the weight of $G = 2750\ kN$ was stacking the material collected from two excavators—Ds-1200 and SchRs-315—during the second shift at the end of October. In the other half of the shift, during normal operation, a slow tilting (as amicably stated by the team) of the bodywork occurred. The titling continued until the end of the stacking extension arm rested on the horizontal heap being stacked. The total bodywork tilting was then approx. 15° (Fig. 13).

This resulted in the bending of one grip, a torn-away upper segment of the ball cradle in the bodywork rotating mechanism and opening of the bodywork within the cradle plane.

During that time, the material load being stacked was located on the horizontal conveyor and belt no. I. Belt no. II and the directly stacking belt no. III were empty. During a visual inspection of the failure site, it was found that the end of the stacking extension arm was covered with material load in a value of approx. 5 m^3, once rested on the heap. Calculated after the failure, the volume of the material load on the horizontal conveyor, given as the output, amounted to $Q_1 = 3794\ m^3/h$, and on belt no. I, the value was $Q_2 = 3553\ m^3/h$. This means that the material feed exceeded the theoretical yield of the stacking machine by 23% (and 15%).

The scope of structure damage was small. The left side of the ball-and-socket joint on the support carriage in belt no. I was torn off, and rivets on the connection on the

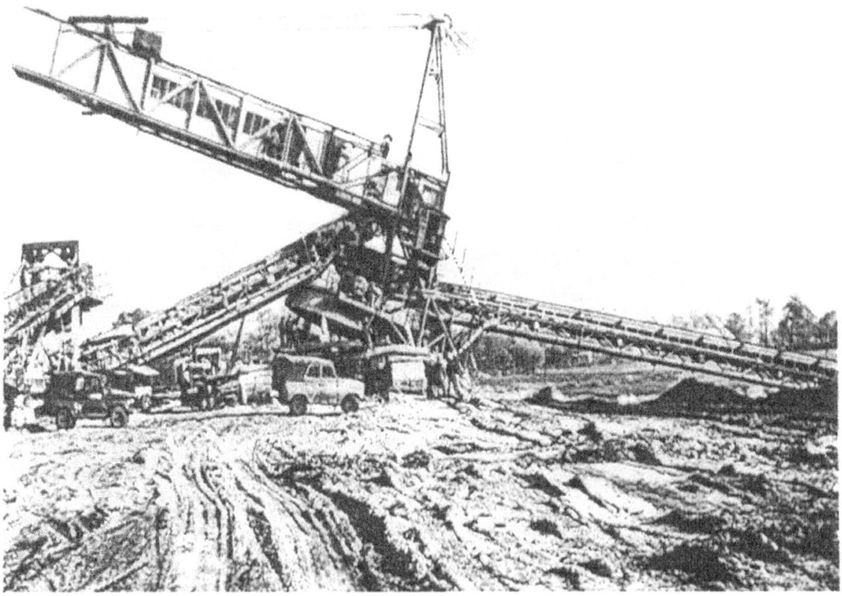

Fig. 13 The stacking machine As-50.12 resting on the heap

lower belt of the load-bearing structure of the stacking extension arm between the first and the second set of suspension ropes were damaged by shearing. Additionally, the transfer hopper from belt no. I to belt no. II was crushed.

During the damage repair, which took 6 days, hook grips were reinforced in addition to repairing the crushed transfer hopper and replacing the segment of the ball-and-socket joint. Causes of the failure may be sought in the obvious mismatch between the stacking machine and the output of the excavators operated alongside. Usually, the maximum output of the belt conveyor collecting the excavated material from excavators is assumed to be higher than the total output of the excavators by 30%, which serves as protection from the effects of bucket overloading while mining light sandy loams of easy workability.

For the described case, outputs of excavators were 1920 m^3/h and 1030 m^3/h, respectively. This means that the total output was 3835 m^3/h (for the opening ratio of 1.3—and exceeded the stacker machine output by 24.5%. This fact probably caused the machine bodywork to lose stability with the resulting failure condition as described. However, there are many questions, and no answer can be found in the available documentation. One of the issues, which is puzzling, is that belts no. II and III were empty during the visual inspection after the failure.

Why, if belt no. I was filled did the machine operate and turn over by itself? What was the actual cause resulting in the lost stability of the bodywork? Surely, the bodywork must have been rather balanced. For empty belts, the plan view of the centre of gravity of the machine bodywork is located inside the perimeter of the ball cradle. Then, the grips are not loaded. In addition, it is not clear why the end of the stacking extension arm was covered with the material. If belt no. III was empty, what was the cause? It surely did not sink by itself. Unfortunately, no one investigated the case in-depth, and it has been passed over in silence for so long that it is impossible to reconstruct the actual causes of the failure.

3.6 Excavator SchRs-350—Broken Loading Extension Arm

A 22-year-old excavator with a theoretical yield of $Q_{theor} = 1750$ m^3/h and weight of G = 12.44 MN was operated to collect the material load (Fig. 14). In the other half of February, at the beginning of the morning shift, a large mound of excessive load entered the hopper at the loading extension arm and clogged it, resulting immediately in overfilling on the loading belt. Similar cases had occurred earlier and excavator personnel had developed their own method of cleaning the hopper.

For that purpose, the loading extension arm was lifted to a height of approx. 6 m, a wooden post was placed under the mound, and the extension arm was lowered to push out the mound upwards. Additionally, this time similar actions were taken. Unfortunately, the upper end of the post slid down unintentionally from the mound

Fig. 14 Excavator SchRs-350. Condition before the failure

and supported the load-bearing structure of the extension arm. The extension arm was lowered by the operator sitting at the control panel on a metal stool without shock absorbers. The operator was commanded by the excavator face man standing at the working level. In this excavator, the loading extension arm was lowered by means of a screw-type hoist equipped with a special T-bar with a nut. The self-aligning support of the extension arm was designed with a clevis pin joint on the structure. As load statics of the extension arm concluded that its structure was always compressed, the joint was designed so that it rested on the half-sleeve, not the pin.

The foreman, who was not able to see the results of the operations being performed, believed that the mound was firmly set in the hopper and ordered the continuation of the lowering of the extension arm. The increasing load on the wooden post during the operation (weight of the extension arm and the excavated material from the transfer point) ultimately caused it to collapse rapidly. Falling under the load, the extension arm was suddenly stopped by the screw mechanism, and the resulting dynamic response caused (Fig. 15):

- bending and breaking of the extension arm structure;
- the joint of the extension arm falling out the half-sleeve in the structure;
- falling of the broken extension arm together with the cabin onto the working level.

Fig. 15 Excavator SchRs-350. Condition after the failure

The damage repair of the excavator took 28 days. The extension arm operator, sitting in the cabin on a rigid metal stool, suffered serious internal injuries.

References

1. Babiarz, S., Dudek, D.: Kronika awarii i katastrof maszyn podstawowych w polskim górnictwie odkrywkowym. (in Polish). Oficyna Wydawnicza Politechniki Wrocławskiej. Wrocław (2007)
2. Zio, E.: Reliability engineering: Old problems and new challenges. Rel. Eng. Syst. Safety **94**(2), 125–141 (2009)
3. Cimellaro, G.P., Reinhorn, A.M., Bruneau, M.: Framework for analytical quantification of disaster resilience. Eng. Struct. **32**(11), 3639–3649 (2010)
4. Moteff, J.D.: Critical Infrastructure Resilience: The Evolution of Policy and Programs and Issues for Congress. Congressional Res, Service (2012)
5. Fang, Y.-P., Pedroni, N., Zio, E.: Resilience-based component importance measures for critical infrastructure network systems. IEEE Trans. Reliab. Inst. Electr. Electron. Eng. (2016)
6. Nowakowski, T., Tubis, A., Werbińska-Wojciechowska, S.: Assessing the supply chain resilience—case study of footwear retail distribution network. Article prepared for European Safety and Reliability Conference ESREL 2016, 25–29 September, 2016, Glasgow (2016)
7. Nowakowski, T.: Vulnerability versus dependability of logistic systems. Proceedings of Carpathian Logistics Congress CLC 2013, 9–11 December 2013, Cracow, Poland (2013)
8. Chlebus, M., Nowakowski, T., Werbińska-Wojciechowska, S.: Supply Chain vulnerability assessment methods—possibilities and limitations. In: Safety and Reliability Of Complex Engi-

neered Systems: Proceedings of the 25th European Safety and Reliability Conference, ESREL 2015, Zurich, Switzerland, 7–10 September 2015 (pp. 1667–1678). CRC Press/Balkema
9. Barker, K., Ramirez-Marquez, J.E., Rocco, C.M.: Resilience-based network component importance measures. Reliab. Eng. Syst. Saf. **117**, 89–97 (2013)
10. Cox, A., Prager, F., Rose, A.: Transportation security and the role of resilience: a foundation for operational metrics. Transp. Policy **18**, 307–317 (2011)
11. Fakoor, A.M., Olfat, L., Feizi, K., Amiri, M.: A method for measuring supply chain resilience in the automobile industry. J. Basic Appl. Sci. Res. **3**(2), 537–544 (2013)
12. Henry, D., Ramirez-Marquez, J.E.: Generic metrics and quantitative approaches for system resilience as a function of time. Reliab. Eng. Syst. Saf. **99**, 114–122 (2012)
13. Francis, R., Bekera, B.: A metric and frameworks for resilience analysis of engineered and infrastructure systems. Reliab. Eng. Syst. Saf. **121**, 90–103 (2014)
14. Singhal, P., Agarwal, G., Lal, Mittal M.: Supply chain risk management: review, classification and future research directions. Int. J. Bus. Sci. Appl. Manag. **6**(3), 15–42 (2011)
15. Supply chain vulnerability. Executive Report, Cranfield University School of Management (2002). http://www.som.cranfield.ac.uk/som/dinamic-content/research/lscm/downloads/Vulnerability_report.pdf
16. Sheffi, Y., Rice, J.B.: A Supply Chain View of Resilient Enterprise. MIT Sloan Manag. Rev. **47** (1) (2005)
17. Sheffi, Y.: The Resilience Enterprise. Overcoming Vulnerability for Competitive Advantage. The MIT Press (2007)
18. Młyńczak, M.: Bucket wheel excavators: past to present experiences in safety operation. In: Proceedings of Probabilistic Safety Assessment and Management PSAM 12, June 2014, Honolulu, Hawaii (2014)
19. Sokolski, P.: On wear processes in pin joints in caterpillars of large-size working machines. In: Fuis, V. (ed.) Proceeding of 23rd International Conference on Engineering Mechanics 2017, pp. 910–913. Brno University of Technology

Overview of Materials Testing of Brown-Coal Mining Machines (Years 1985–2017)

Łukasz Konat and Grzegorz Pękalski

Abstract This multi-subject study is focused on some crucial material issues related to brown-coal mining machines. These issues include degradation problems of rimmed steels used till the 1980s for structures of excavator bodies. The performed research works resulted in resigning this group of materials in construction of new or modernised machines. Degradation examinations and the resulting evaluation of degradation condition of the materials are currently applied for bridge structures. Systematic selection of materials (together with complex heat treatment) can be illustrated by the presented process of selecting materials for the travelling unit of a machine. Problems of abrasive wear under dynamic loads are presented on the examples of bucket teeth of excavators, pad-welded layers of chutes of a bucket wheel and low-alloy martensitic steels (with Hardox steels especially considered). The main thesis of the presented study, documented also by results of a service experiment, is that wider application of these materials can be critically important for the service life of opencast mining machines. The main issue is elimination, by proper thermal treatment, of heat-affected zones in the vicinity of welded joints, characterised by reduced resistance to abrasive wear. A thermal treatment of welded joints is also suggested, leading to homogenisation of microstructures in heat-affected zones and in base materials, as well as to definite improvement of mechanical properties of the joints.

Keywords Mining machines · Material issues · Degradation · Travelling mechanism · Abrasive wear · Low-alloy martensitic steels · Welded joints

1 Introduction

Comprehensive research works related to material issues in surface-mining machines were initiated at Mechanical Faculty of Wroclaw University of Science and Technology in the mid-80s of the twentieth century. Previously (e.g. in the 60s of the twentieth

Ł. Konat (✉) · G. Pękalski
Faculty of Mechanical Engineering, Wroclaw University of Science and Technology, Wybrzeze Wyspianskiego (ul.) 27, 50-370 Wroclaw, Poland
e-mail: lukasz.konat@pwr.edu.pl

M. Sokolski (ed.), *Mining Machines and Earth-Moving Equipment*,
https://doi.org/10.1007/978-3-030-25478-0_2

century), numerous expert opinions were elaborated about this topic. The expertises, elaborated by Janusz Zipser and Rudolf Haimann, concerned e.g. testing of materials used for teeth of excavator buckets. Wider studies undertaken as a response to the demand of the industry were based on the literature items [1–3]. The books contain information (and even separate chapters) concerning material problems related to complex constructional, technological, mechanical and geological issues. A value of these publications is also the fact that they treat the analysed engineering objects in holistic terms, indicating a necessity to consider constructional, technological, mechanical and geological aspects at the same time, considering their interactions. The relationship taken from [4] and interpreted in details in [5], shown in the diagram below (Fig. 1), very well illustrates the present-day systematic approach to the problems of degradation (including its structural aspects).

A consequence of such a way of operation was developing a complex plan of examinations [6] that would make it possible to identify condition of the object, see Fig. 2. The specified examinations were carried out in a wide scope for the body of an excavator SchRs 1200 that was in operation since 1943 (operation period 55 years). Eight constructional joints were selected on the grounds of mechanical calculations and analyses of the research questionnaires obtained from a dozen of prominent experts, users of the examined machines. At the end of 90s of the twentieth century, not all of the examinations mentioned in Fig. 2 were executed. It is only in the first decade of twenty-first century that the works concerning the problem of cracking in relation to microstructure of the materials under dynamic loads were undertaken, e.g. [7].

A significant role in the decision on leaving a machine in operation, its modernisation or withdrawal from operation played results of the so-called 'laboratory strategy'. Indeed, the decision on modernisation had to include suggestions of new material solutions, documented by results of examinations. The works within the

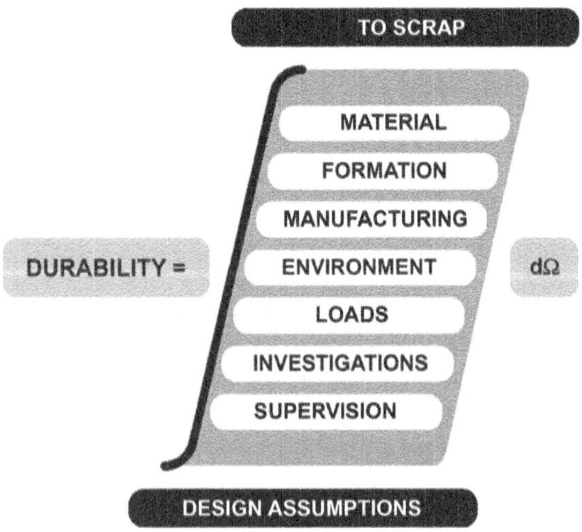

Fig. 1 Schematic presentation of durability of technical objects as integrand of individual durability-related factors [4]

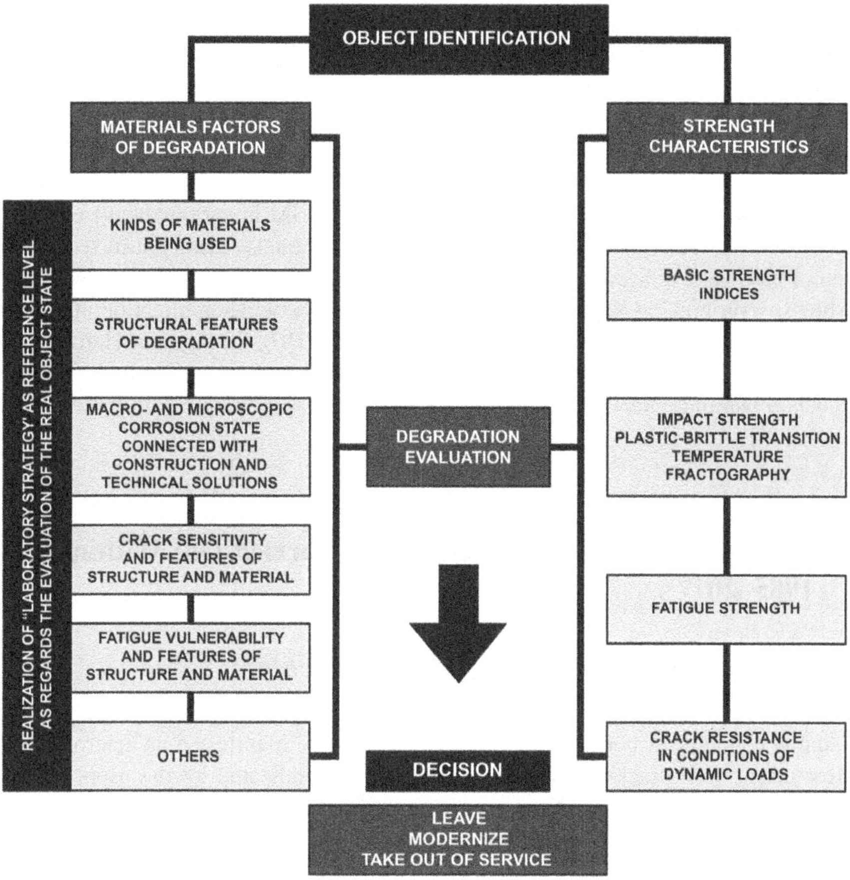

Fig. 2 Schematic presentation of multicriterial system for solving material problems in the degradation theory [6]

'laboratory strategy' concerned e.g. evaluation of corrosion resistance of 'old and suggested new materials' (e.g. [8]), evaluation of susceptibility of new materials to microstructural degradational changes (e.g. [6]) or abrasive-wear resistance (e.g. [9]). In connection with the implementation of the concept to join constructional components by welding methods [10], an urgent need appeared to examine welded joints of low-alloyed higher-strength steels and of low-alloyed martensitic steels [11, 12]. Several-years' works on the research program shown in Fig. 2 made it possible to create a comprehensive database concerning symptoms, course and measures of degradational processes. A large set of comparative data concerning suggestions of new material solutions was also obtained. Next, that made it possible to develop a schedule of degradational examinations adapted to the operators' needs. This schedule permits the degradation condition to be assessed with no necessity to put the machine out of action and restricts scopes of research works and methods of testing [13].

It can be emphasised that results of material examinations of surface-mining machines carried-out in harmonious co-operation with their designers and users were successively implemented. The following examples can be mentioned: withdrawal from using steels susceptible to structural degradation processes for bodies of the machines, wide application of cast steel L35GSM in various heat-treatment conditions for track links, drivers of driving wheels and bucket teeth, optimised selection of pad-welded layers on chutes of bucket wheels. However, it seems that more expressive progress at solving material problems will be wider application of highly abrasion-resistant martensitic steels (excavator KWK 910). Steel grades Hardox 400 and Hardox 500 were experimentally verified as lining plates on chutes of bucket wheels, obtaining excellent results in comparison to steel plates with pad-welded layers [14].

2 Selected Results of Research Works Carried-Out Within 1985–2017

2.1 Construction of Bodies of Mining Machines

Usually, damages of bodies of mining machines were manifested as macrocracks. Most often, such cracks were located in the joints indicated by the users on the grounds of their service experiences and established by strength calculations (critically stressed joints). When overloads or random events are omitted, the causes of the cracks are microstructural degradational changes or corrosive processes. Results of numerous research works (more extensively presented in [5]) indicate that, as a result of long-lasting dynamic loads of steels, precipitation of iron carbides and nitrides proceeds, accompanied by decomposition of (sparse) pearlite areas. Extreme degradation condition of microstructure is shown in Fig. 3. Such processes proceed in especially intensive way in low-carbon rimming steels. Microstructure changes can result in increased yield strength and tensile strength, missing physical yield point, decreased impact strength and increased brittle fracture transition temperature, see Fig. 4.

The so-far published results (e.g. [6]) indicate that to determine the susceptibility of steel to degradation processes, it is enough to do the following:

- to determine the chemical composition of steel, limited to five basic elements: C, P, S, Mn, Si;
- to carry out observations by means of light microscopy;
- to determine curves of impact strength versus temperature in post-service and normalised conditions.

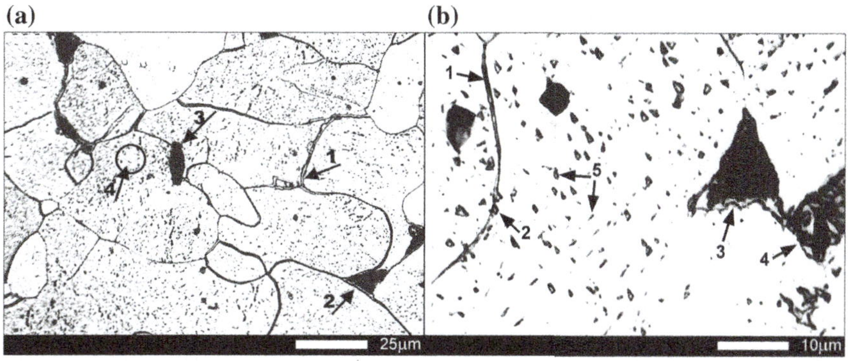

Fig. 3 Microstructure of steel with chemical composition corresponding to St3SX from a fragment of the bucket wheel; the material is ca. 55 years old [5]: **a** ferrite grains with non-continuous envelope of cementite (1), small pearlite areas (2) and non-metallic inclusions (3), numerous iron carbides and nitrides inside ferrite grains (4); **b** magnified fragment of microstructure shown in Fig. 3a. Two ferrite grains are separated by the boundary decorated with non-continuous envelope of tertiary cementite (1) and nitrides (2). Visible are also two areas of fine-lamellar pearlite (3) and pearlite with features of decomposition (degenerated) (4). Numerous particles of cementite and iron nitrides inside ferrite grains (5). Etched with 2% HNO_3, light microscopy

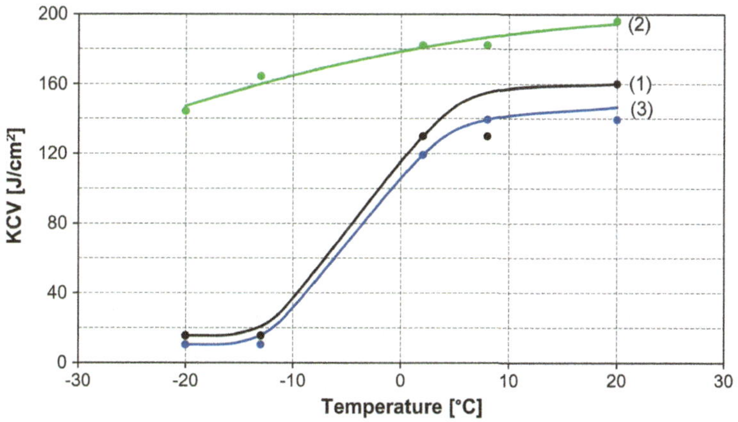

Fig. 4 Changes of impact strength of steel St3SX after 55 years of service [15]: curve 1—post-service condition, curve 2—after normalisation at 930 °C for 1 h, curve 3—after normalisation and ageing at 200 °C for 1 h

Such a reduction of the examination scope results from the repeatedly confirmed facts that, in highly degraded steels, physical yield point also occurs with respect to low carbon content. However, it should be noted that this reduction (even in relation to the scope of examinations adapted to the operators' needs, shown in [13]) is possible only when a very wide comparative database is available.

Another factor deciding the condition of a worn-out structure of an excavator body is corrosive processes. In relation to the machines operated for several dozen years, the following factors conducive to development of corrosion can be indicated [16]:

- design solutions (angles, channels, flat bars) creating 'corrosive constructional traps' in node points, see Fig. 5;
- joining technology (riveted joints) and repair technology (welded joints) conducive to crevice corrosion. Moreover, there is a risk that acicular structures with larger specific volume can be formed in heat-affected zones of welded joints and the emerging stresses summing-up with thermal stresses can initiate cracks. It should be also emphasised that precipitation processes are intensified in heat-affected zones, and areas with degraded microstructure are created in course of long-lasting service, see Fig. 6;
- small resistance of low-carbon rimmed steels to atmospheric corrosion.

These days, risks resulting from structural degradation and corrosion are to a large extent eliminated. This results from understanding the above-described dangers and their results. As for constructions of modernised and new excavator bodies, steels susceptible to structural degradation (cheaper) were resigned in favour of higher-strength steels based on the grades 18G2 or 18G2A. As was demonstrated in [6], these steels do not show susceptibility to microstructural degradation, at the same time revealing higher resistance to atmospheric corrosion. A change of the body design and withdrawal from using structural shapes in favour of flat bars joined by welding technologies [10]. In this question, beneficial is also reduction of environment pollution, e.g. in the so-called 'Black Triangle'. For example, in the period from 1990 to 2013, the power plant 'Turów' reduced emission of SO_2 by 84% and emission of dusts by as much as 98% [17].

Fig. 5 Macroscopic images of a riveted and a welded joint of bucket wheel jib of an excavator SchRs-1200 (joint 25) [5]: **a** general view of the joint with macroscopic damages like craters, discontinuous weld or unfilled weld close to the fracture surface; **b** magnified fragment of the joint shown in Fig. 5a in fracture area (upper part)—corrosive gaps (riveted and welded joints) with developed crevice corrosion are marked with arrows 1–4

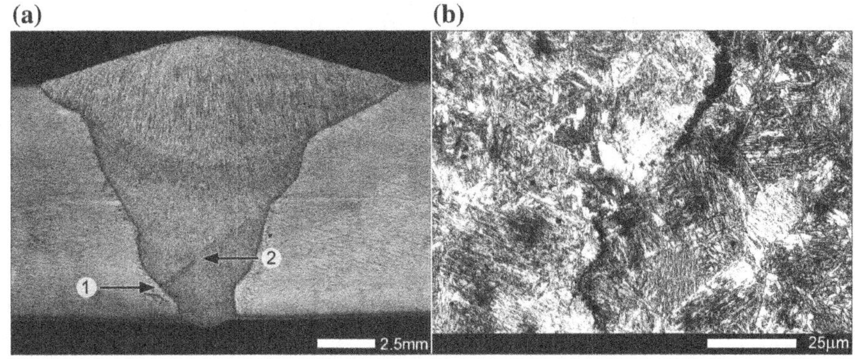

Fig. 6 Exemplary welded joint with a crack in heat-affected zone, initiated by thermal welding processes [5]: **a** macroscopic view of weld with a crack running from fusion line (1) deep into deposited material (2); **b** microstructure in cracking zone—widespread weld crack filled with corrosion products against a background of matrix material with acicular morphology. Etched with 2% HNO_3, light microscopy

The idea and practice of studies on the processes of steel degradation after long-term service are currently used at evaluation of wear condition of structures of old bridges and overbridges [18]. This especially concerns bridges and overbridges of Wrocław, whose structures were made of puddled steels (manufactured till 1905). At present, evaluation procedures of such objects include the item 'evaluation of material degradation degree'. The 'degradation database' was also significantly completed with structural components of the main railway station in Wroclaw that was revitalised in the years 2010–2012 [19]. So, it can be said that the degradation theory—in the discussed case, material issues of the degradation theory—found a valuable practical application.

2.2 Traversing Unit of the Excavator (Assembly Track Link—Crawler Tread)

Research works related to traversing units of brown-coal excavators were initiated by an expertise in 1990, aimed at determining causes of cracking of welded joints between links made of cast steel L35GSM and crawler treads of steel St3S or 18G2. Figure 7 shows layouts of selected constructional links of traversing unit of a brown-coal excavator. Cracking zones of welded joints are marked with a broken line (18).

It was relatively easy to establish causes of cracking of welded joints, since it was reduced to finding presence of hard and susceptible to cracking martensitic microstructures in the heat-affected zone (from the side of cast steel L35GSM). For welded joints L35GSM-18G2A, typical heat-treatment operations were applied, consisting of normalising and stress-relief annealing. Each of these operations was

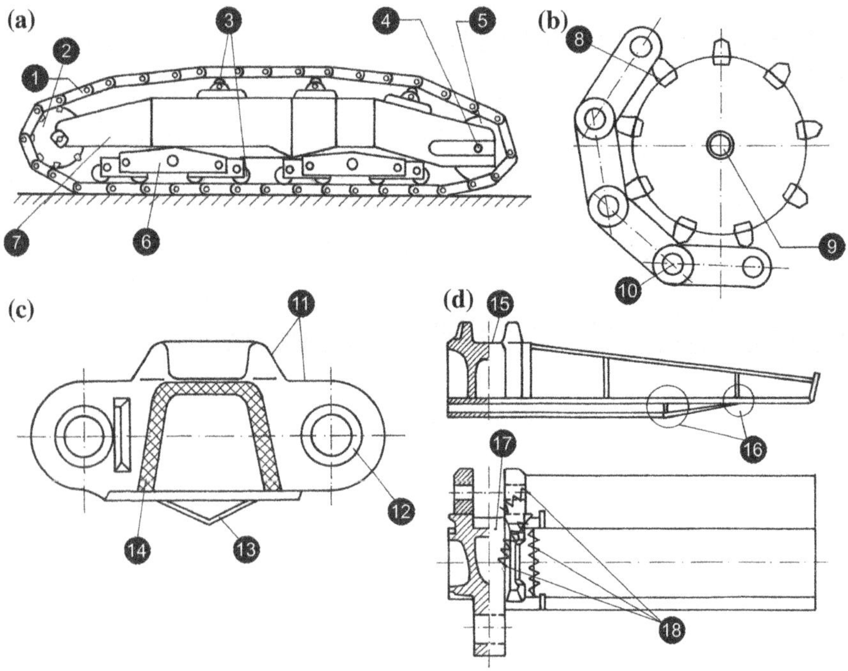

Fig. 7 Layouts of selected constructional links of traversing unit of a brown-coal excavator: **a** general layout of the traversing unit; **b** schematic mating of driver wheel with track links; **c** layout of track link; **d** layout of crawler tread; 1—crawler tread, 2—driver wheel with drivers, 3—track wheels, 4—idler wheel shaft, 5—idler wheel, 6—rocker unit, 7—girder unit, 8—mating surface wheel-driver, 9—driver wheel shaft, 10—pin of track link, 11—mating surfaces with drivers, 12—link sleeve, 13—spur, 14—welded area cast steel-steel, 15—mating surface with track wheels, 16—welded area steel-steel, 17—link raceway, 18—cracking zone of welded joints

effective, reducing maximum hardness of the joint below 350 HV and favourably changing its microstructure, see Fig. 8.

Therefore, the problem could be found solved, if not the trouble with executing heat treatment on a real object. A necessity appeared to purchase and start up a suitably big furnace in that six assemblies track link—crawler tread could be placed. The idea of normalizing had to be rejected with respect to geometrical deformation of the assembly at the required temperature of 930 °C. So, after welding, the assembly link-tread was subjected to stress-relief annealing at 650 °C for 5 h. Execution of the above-described works inspired the group of designers and users of these machines, as well as employees of Wroclaw University of Science and Technology to take up wider research works on the entire traversing unit. The performed analyses of damages of this mechanism showed that the track link is a vulnerable point—cracking of these links was often found, see Fig. 9.

Another cause was incorrect material microstructure (Fig. 10) resulting in expanding the mating surface between links and rollers or between link runners and drivers.

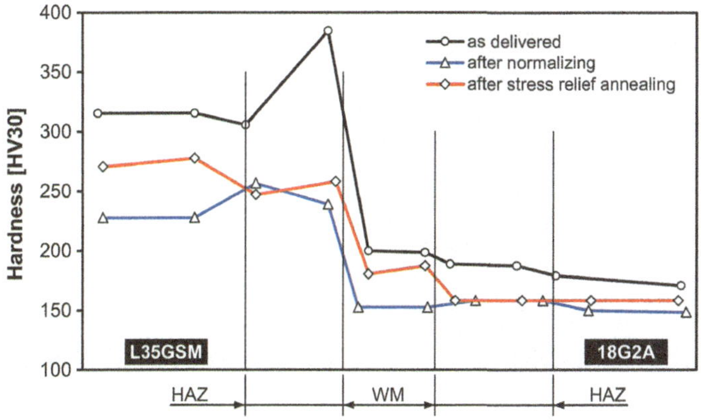

Fig. 8 Hardness of welded joint of cast steel L35GSM with steel 18G2A [20, 21]: HAZ—heat-affected zone, WM—weld material zone

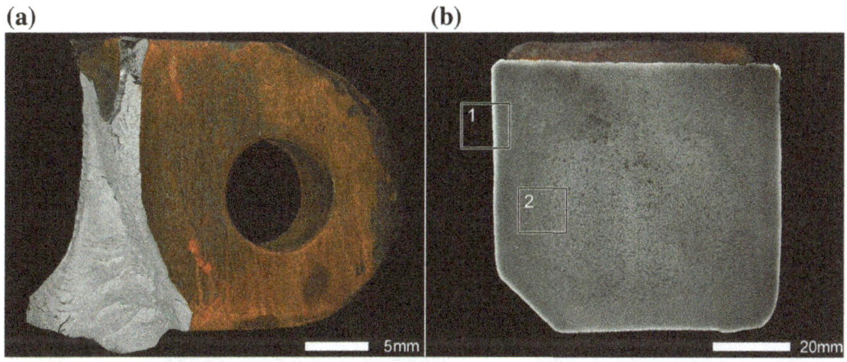

Fig. 9 Macroscopic images of eye of a track link made of L35GSM: **a** fracture surface of cleavable nature with numerous leaps and macrocracks; **b** macrostructure on cross-section of the eye. Unetched (**a**) and etched with 2% HNO_3 (**b**), stereoscopic microscopy

That led to the disappearance of the space between link eyes in the crawler chain, resulting in faulting of crawler treads and making movement of the excavator impossible. During the performed examinations [22], it was additionally demonstrated that, in many cases, cracking of the eyes was caused by application of cast steel L35GSM directly in the 'as cast' condition, see Fig. 11. It can be said that this condition is similar to normalised condition resulting after air cooling of castings. Trials of quenching and tempering the links according to the parameters given in PN-88/H-83160 were also carried out. However, the experiments carried out at the smelter 'Huta Małapanew' in Ozimek did not deliver any satisfactory results.

The above-presented statements brought to a research program that was finished with elaboration and implementation of complex heat treatment of track links and entire link-tread assemblies after welding. With regard to the necessity of machining,

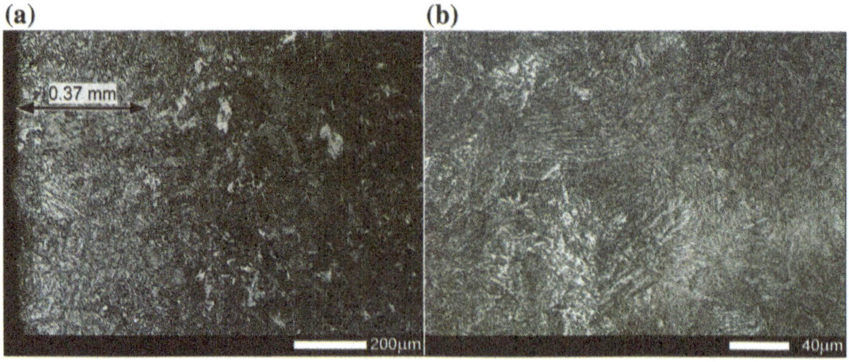

Fig. 10 Microstructure of L35GSM after incorrect heat treatment—sample taken from the eye shown in Fig. 9: **a** sub-surface area with marked decarbonised zone (Fig. 9b—frame 1)—non-equilibrium acicular structure with ferrite areas; **b** magnified image of the area marked with frame 2 in Fig. 9b—martensitic-bainitic structure. Etched with 2% HNO_3 (**b**), stereoscopic microscopy

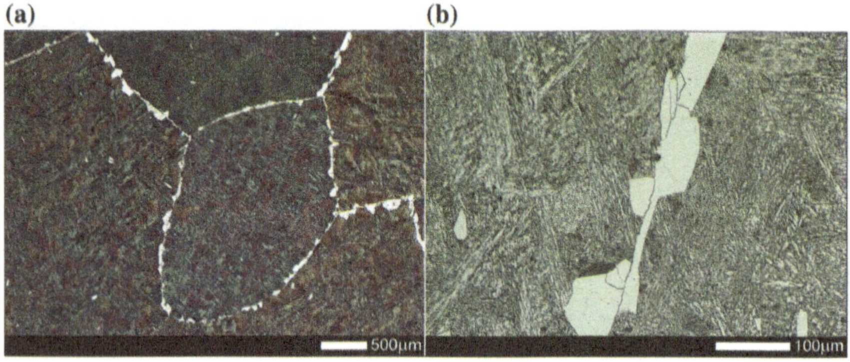

Fig. 11 Microstructure of L35GSM in as-cast condition: **a** coarse-grained structure of non-equilibrium pearlite with bright envelope of ferrite on grain boundaries; **b** magnified view of microstructure shown in Fig. 11a—bright ferrite areas making envelope on grain boundaries of non-equilibrium pearlite. Single particles of ferrite inside grains. Etched with 2% HNO_3, light microscopy

the start point was the assumption that temper sorbite would be obtained in the link material (on possibly largest area), with hardness not over 260 HB. This structure (Fig. 12) is characterised by optimum ratio of yield point to tensile strength and makes the best 'substrate layer' for the microstructures obtained by surface hardening. It was recognised that mating surfaces of links with rolls and drivers should have the same structures and much higher hardness. This resulted in changing the driver's material (also to L35GSM) that was heat-treated together with track links.

In [23, 24], comprehensive research works concerning of final heat treatment of the mentioned constructional unit of the excavator are described in details. In brief, this heat treatment can be presented as follows:

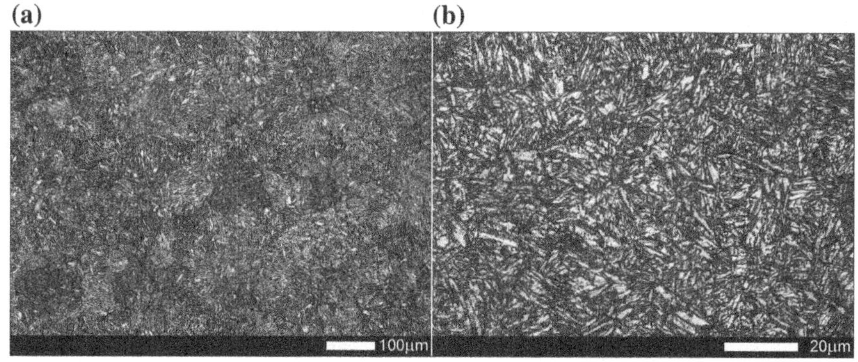

Fig. 12 Exemplary microstructure of correctly heat-treated cast steel L35GSM: **a** water quenched after soaking at 930 °C for 1 h and tempered at 600 °C for 2 h followed by water cooling; **b** magnified fragment of structure shown in Fig. 12a—temper sorbite with post-martensitic orientation. Etched with 2% HNO_3 light microscopy

- casting track links (weight from 350 to 500 kg) with subsequent air cooling. Inspection of chemical composition and mechanical properties on test rods. An important element of the process is controlling the concentration of Mo at minimum 0.3% to prevent temper brittleness and to obtain impact strength over 26 J/cm^2;
- slow heating to hardening temperature 930 °C, soaking for the time-dependent on the link mass and quenching in water;
- tempering at 600 °C for the time-dependent on the link mass and cooling in water. Soaking times are selected (properly short) to ensure maintaining the hardening structures. The same heat-treatment operations would be also applied to drivers;
- welding track links with crawler treads—preliminary heating of links in the area two times wider than the expected heat-affected zone and slow cooling after welding;
- stress-relief annealing of the assembly link-tread at 650 °C for 5 h. This operation causes relaxation of stresses in a welded joint and changes microstructure of heat-affected zone. As a result, microstructure of temper sorbite with post-martensitic orientation is obtained, characterised by hardness ca. 260 HB and impact strength ca. 100 J/cm^2;
- surface hardening of mating areas between links and rollers or drivers with no separate operation of low tempering. The heat accumulated in the link material after stress-relief annealing is sufficient for spontaneous 'self-tempering'.

The basic problem with execution of the presented heat treatment of track link assemblies was its long time. Execution of all stages reached 40 h, which was seriously questionable from the economical point of view. Nevertheless, the results of service trials performed at the mine Turów could convince users of brown-coal mining machines to implement the suggested technology.

A material used for track links, competitive to L35GSM, is high-manganese austenitic cast steel L120G13 containing ca. 1% C and 13% Mn. Its application

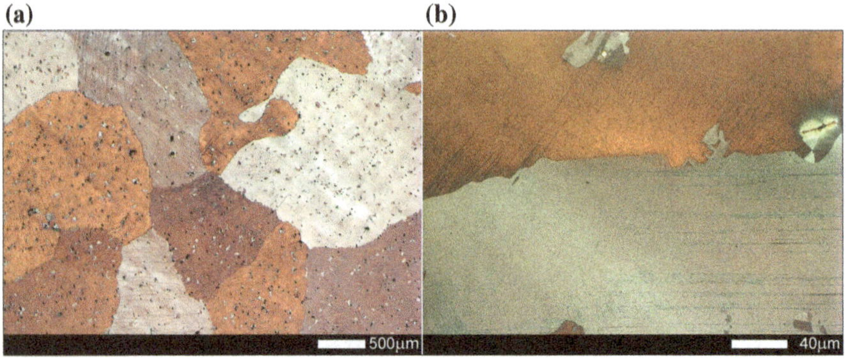

Fig. 13 Microstructure of cast steel L120G13: **a** grains of alloyed austenite with numerous carbides inside grains and locally on their boundaries; **b** magnified microstructure of L120G13 after solution heat treatment—no carbide phases inside and on grain boundaries of alloyed austenite. Etched with Mi17Cu (according to PN-H-04503:1961 P), light microscopy

for selected parts of basic brown-coal mining machines is always connected with problems at heat treatment of track links, resulting in cracking of eyes. The problems of heat treatment of cast steel L120G13 are mainly related to a continuous envelope of manganese cementite $(Fe, Mn)_3C$ present in the structure on grain boundaries of alloyed austenite, see Fig. 13).

Currently, track links are basically made of cast steel L35GSM (heat-treated acc. to the above-presented principles) or of high-manganese austenitic cast steel L120G13 with an addition of titanium.

2.3 *Components and Units Subjected to Abrasive Wear Under Dynamic Loads*

The questions related to abrasive wear of constructional materials are ones of crucial questions in many fields and industrial sectors. In relation to working machines of surface mining, randomness and large amplitudes of loads occurring at abrasive conditions should be always taken into account. Among components and units of these machines subjected to such loads, the following can be mentioned: bucket teeth, bucket blades, chutes, hoppers, transfer stations and slides (e.g. of a bucket wheel). With respect to materials, the problems can be solved in various ways, for example:

- by using materials with high abrasion resistance that, in fact, not necessarily demonstrate this feature. These materials include e.g. steel 35SG and austenitic cast steel applied in the hope that the material will be strengthened (hardened) by plastic deformations. This group includes also, although still in a very limited range, low-alloy martensitic steels (e.g. Hardox).

- by using materials showing higher abrasion resistance after special technological operations applied exclusively to the most heavily loaded layers (cross-sections). This group of solutions includes bucket teeth WKL3 cast of L35GSM and heat-treated. In addition, quenching and low tempering is applied to these elements, within the cutting areas only. Trials of increasing abrasion resistance of the teeth were also carried-out by directional remelting with electron beam, by fitting inserts made of sintered carbides or by using cast layers with modified chemical composition.

However, the most often used solution is imposition of pad-welded layers on surfaces exposed to abrasive wear. This technology has many advantages since the market offers a great number of pad-welding electrodes that make it possible to adapt properties of the imposed layers to the occurring loads and geological conditions. At present, pad-welding processes are robotised, which permits even (free of seams) pad-welded surfaces to be obtained on large areas. This significantly affects hardness of lining plates of chutes and transfer stations. Finally, pad welding is an effective and technologically easy operation of repair and reclamation.

The authors of this study directly participated in examinations of bucket teeth WKL3 and bucket corners [25, 26] and conducted research works related to ca. 50 types of pad-welding electrodes, e.g. [27–30]. Analysis of literature data and own results makes it possible to formulate some general statements in relation to pad-welded layers. The first generalisation concerns incorrect approach to the questions of abrasive wear. It comes from simplified belief about linear relations between hardness and resistance to abrasive wear. In that connection, there is still a tendency among manufacturers of pad-welding electrodes to introduce to their chemical composition high percentages of strongly carbide-forming elements creating very hard simple carbides. For example, some of the examined pad-welded layers included 9% Mo (CastalinXHD-6715), 10% V (EP3600H24V9), 9% Nb (CastalinXHD-6715) and 2.8% B (PD9). Another error is omitting or too superficial treating the issues of macro- and microstructures of pad-welded layers. These structures are various on the layer thickness, which basically affects prediction of their durability. Manufacturers of electrodes usually give out, apart from chemical composition, their hardness measured generally on the surfaces, e.g. 62 HRC for padding weld PD9. Figure 14 shows exemplary macro- and microstructures of a selected pad-welded layer with highly variable structural features confirmed by hardness measurements, see Fig. 15.

The additional factors, strongly affecting the possibility of predicting lifetimes of selected machine parts, are characteristics of wear of pad-welded layers. It was shown that the influence of macro- and microscopic defects and condition of fusion zone can prevail over chemical composition of the electrode, correctly matched to the service conditions. A fragment of a chute lined with pad-welded plates is shown in Fig. 16.

As can be seen in Fig. 16, a part only of lining plates is worn by abrasion. In most cases, abrasive wear of pad-welded plates is decisively affected by detaching from the substrate or surface spalling of the padding weld.

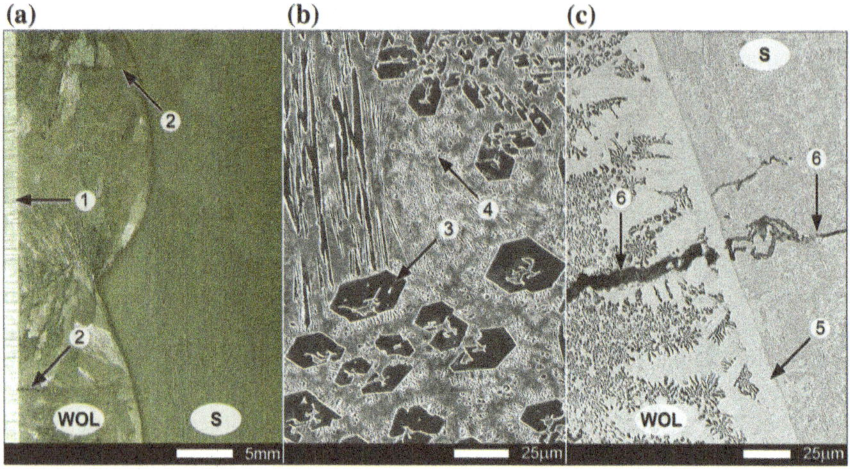

Fig. 14 Structures of a layer pad-welded with electrode Most EL-Hard 70 [30]: **a** macroscopic view of pad-welded layer. At the left, grinding marks (1) and two wide cracks (2) are visible in the weld material; **b** microstructure in padding weld zone. Primary chromium carbides (3) and secondary boron carbides (4) in a mixture of alloyed ferrite and carbides; **c** microscopic view of fusion zone. Clearly visible 'strip' of alloyed ferrite (5) and wide microcrack (6) present also in substrate material. WOL—pad-welded layer, S—substrate material. Etched with 3% HNO_3 and electrolytically with chromic acid, light microscopy (**a**) and SEM (**b** and **c**)

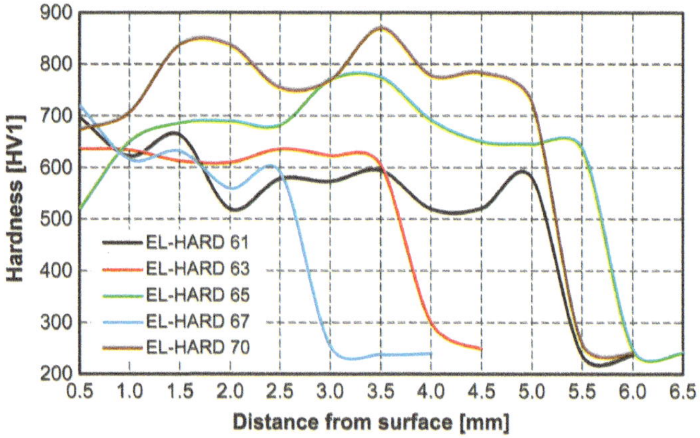

Fig. 15 Exemplary hardness values on cross-sections of layers pad-welded with electrodes Most EL-Hard [30]

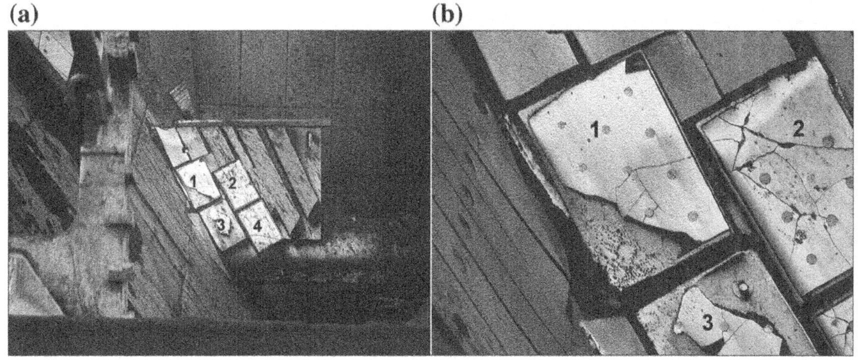

Fig. 16 Fragment of chute of bucket wheel of excavator KWK-1500s lined with cast iron plates (1–4) fitted on the lump breaker: **a** general view of the chute; **b** macroscopic view of surfaces of plates pad-welded with Fe Cr–C alloy in post-service condition

2.4 Initial Examinations of Hardox Steel

Research works on low-alloyed martensitic steels, which have been for a dozen years the main subject matter of the team engaged in material issues in surface-mining machinery, were initiated by KWB-Konin (engineers Stanisław Storch and Zbigniew Woźniak).

The considered group of materials includes Hardox steels. On the grounds of characteristics of Hardox 400, published in the manufacturer's materials, the mine 'Konin' purchased material for bucket blades to be used in the research. The blade bars were fitted to the bucket by welding. Results of that experiment were completely negative. The blade made of Hardox 400 was worn in the time twofold shorter than that made of steel 18G2A. Maximum wear areas were observed in heat-affected zones of welded joints. The negative experience with this steel contributed to beginning research works on this group of materials and, especially, on possibilities of using them in surface-mining machines [31].

It was demonstrated in course of these examinations that application of these steels in construction of mining machines can be literally crucial for durability of chutes, hoppers and transfer stations. This thesis was confirmed by a service experiment. With respect to the high importance of this subject, a separate chapter in this publication is dedicated to these steels.

3 Low-Alloy Martensitic Steels

3.1 Introduction

According to available marketing information, low-alloy martensitic steels meet selection criteria and service conditions presented in the considerations above. At present, a great number of steel grades from various manufacturers can be found in the market of the materials defined also as 'abrasive-wear resistant'. The most popular in Europe are [14, 32] Hardox from SSAB-Oxelösund, Raex, B13, B24 and B27 from Rautaruukki (currently also SSAB), 20MCB5, 23MCB5, 28MCB5 and 30MCB5 from Zerneri Acciai SPA, XAR and TBL from ThyssenKrupp Steel Europe AG, Durostat and Brinar from Grobblech GmbH, Fora and Creusabro from Industeel, HTK from Polish Hut-Trans Katowice Spółka z o.o. and many others.

The most popular group of steels with trade name Hardox was introduced to the market in 1970 by Swedish concern SSAB-Oxelösund [33]. According to catalogue data, these steels are characterised by good machinability, very high mechanical parameters, resistance to impact loads and, especially underlined by the manufacturer, good weldability. The main classification criterion of these steels is Brinell hardness. Therefore, the following grades of these steels are distinguished: Hardox 400, Hardox 450, Hardox 500, Hardox 500 Tuf, Hardox 550 and Hardox 600. In addition, apart from the above-presented classification, the company SSAB introduced later the grades Hardox HiTuff, HiTemp and Hardox Extreme, named as the hardest steel sheets in the world [34]. Chemical and mechanical properties of selected Hardox steels are given in Tables 1 and 2 together with their equivalents offered by competitive manufacturers.

Analysis of available information materials on the steels presented in Tables 1 and 2 indicates that all these materials are delivered in hardened condition (quenched in water or oil). An exception can be steel B27 that is also offered after hot rolling, to be heat-treated by the customer. In the case of required hardness level, tempering at 200–700 °C is also carried out. Analysis of chemical compositions indicates that parameters of heat treatment must be diversified, even within one grade of steel. This results mostly from very wide ranges of available sheet thickness. For example, for Hardox 400, available in thickness range between 4 and 130 mm, carbon concentration can range within 0.14–0.32%.

Own chemical analyses (Table 1) confirm this relationship, especially in the context of minimum concentrations of carbon and other elements, no longer published by SSAB for a dozen years. It should be emphasised that the data available for Hardox 400 in the years 1999–2005 covered full ranges of chemical compositions, unlike the currently published maximum values only. In this connection, it can be indicated that concentrations of elements in Hardox 400 change within 0.30–2.50% for chromium, 0.25–1.50% for nickel and 0.25–0.60% for molybdenum. The direct reason for such differentiated chemical composition is a necessity to use similar hardness in the entire cross-section of sheet, resulting also from homogeneous structure of the material. For the wide range of sheet thickness, these properties are realised

Table 1 Chemical compositions of selected low-alloy martensitic steels: MD—manufacturer's data [34–39], OR—own results [12, 14, 27, 31]

Steel grade		# [mm]	C	Mn	Si	P	S	Cr	Ni	Mo	B	CEV
			Selected element—maximum concentration [wt%]									
Hardox HiTuf	MD	40–160	0.20	1.60	0.60	0.050	0.020	0.70	2.00	0.70	0.005	0.56-0.67
Hardox 400	MD	4–130	0.32	1.60	0.70	0.025	0.010	2.50	1.50	0.60	0.004	0.41–0.82
	OR	8	0.12	1.03	0.49	0.013	0.008	0.39	0.07	0.01	0.002	0.38
Hardox 450	MD	3-130	0.26	1.60	0.70	0.025	0.010	1.40	1.50	0.60	0.005	0.48–0.67
	OR	10	0.22	1.09	0.28	0.01	0.000	0.21	0.06	0.13	0.001	0.46
Hardox 500	MD	4–103	0.30	1.60	0.70	0.020	0.010	1.50	1.50	0.60	0.005	0.53–0.75
	OR	10	0.26	0.20	0.75	0.005	0.005	0.70	0.05	0.01	0.001	0.44
Hardox 550	MD	8–65	0.37	1.30	0.50	0.020	0.010	1.40	1.40	0.60	0.004	0.76
Hardox 600	MD	6–65	0.47	1.40	0.70	0.015	0.010	1.20	2.50	0.70	0.005	0.76–0.87
	OR	12	0.44	0.53	0.17	0.006	0.002	0.31	2.03	0.14	0.002	0.76
Hardox extreme	MD	8–19	0.47	1.40	0.50	0.015	0.010	1.20	2.50	0.80	0.005	0.65
	OR	10	0.44	0.49	0.16	0.006	0.002	0.83	2.01	0.14	0.002	0.86
Brinar 400	MD	< 80	0.18	2.00	0.50	0.015	0.005	1.55	NA	0.60	0.005	NA
	OR	12	0.20	1.13	0.23	0.012	0.001	0.61	0.45	0.31	0.002	0.61
Brinar 500	MD	< 60	0.28	1.50	0.80	0.020	0.005	1.50	NA	0.40	NA	NA
	OR	12	0.30	0.97	0.60	0.015	0.001	0.87	0.04	0.20	0.001	0.68
XAR 600	MD	4–50	0.40	1.50	0.80	0.025	0.010	1.50	1.50	0.50	0.005	0.79
	OR	10	0.37	0.85	0.19	0.014	0.001	0.83	1.21	0.15	0.002	0.79
B27	MD	2.5–80	0.32	1.50	0.40	0.020	0.015	0.60	NA	NA	0.005	0.54
	OR	10	0.28	1.26	0.23	0.009	0.006	0.32	0.05		0.002	0.56
HTH 900H	MD	3–50	0.18	1.50	0.45	0.025	0.010	1.00	0.30	0.40	NA	0.48–0.58
	OR	8	0.09	1.41	0.33	0.009	0.005	0.03	0.11	0.03	0.001	0.34

CEV = C + Mn/6 + (Cr + Mo + V)/5 + (Cu + Ni)/15; # sheet gauge; *NA* no available data

Table 2 Mechanical properties of selected low-alloy martensitic steels [34–39]

Steel grade		$R_{p0.2}$	R_m	A_5	KCV^L	KCV^T	HBW
		[MPa]	[MPa]	[%]	[J/cm^2]	[J/cm^2]	
Hardox HiTuf	40–160	850–950†	900–980†	16†	88 (−40 °C)†	50 (−40 °C)	310–370
Hardox 400	4–130	900–1100	1250	10	56 (−40 °C)	NA	370–430
Hardox 450	3–130	1000–1300	1400	10	63 (−40 °C)	NA	390–475
Hardox 500	4–103	1250–1400	1550	8	46 (−40 °C)	NA	450–540
Hardox 550	8–65	1400	1700	7	38 (−40 °C)	NA	525–575
Hardox 600	6–65	1650	2000	7	25 (−40 °C)	NA	550–640
Hardox extreme	8–19	NA	NA	NA	NA	NA	650–700
Brinar 400	<80	900	1200	12	25 (−20 °C)	NA	340–440
Brinar 500	<60	1350	1500	8	25 (−20 °C)	NA	480
XAR 600	4–50	1700	2000	8	25 (−20 °C)	NA	>550
B27	2.5–80	1200	1600	6	NA	NA	470
HTK 900H	3–50	1000	1250	10	NA	NA	380–420

sheet thickness; *L* specimen axis parallel to rolling direction; *T* specimen axis perpendicular to rolling direction; † manufacturer's data from 2005 to 2010, currently unavailable; *NA* no available data

by proper level of hardenability obtained both by addition of 0.002–0.005% of boron and by increased concentrations of Cr, Ni and Mn. In the case of low-alloy steels, conceivable is also—purposeful or accompanying the welding processes—tempering operations at 250–400 °C, which can result in irreversible temper brittleness. This phenomenon is most often prevented by addition of molybdenum over 0.20%. Therefore, big variability of chemical composition of Hardox steels, as well as of other low-alloy abrasion-resistant steels, must be related to high degrees of complexity and individualisation of heat-treatment processes, and must result in differentiated final structures of steels. In Hardox steels, martensitic-type structures and structures with post-martensite orientation created during tempering can be expected. Figure 17 shows exemplary structures of selected steel listed in Tables 1 and 2.

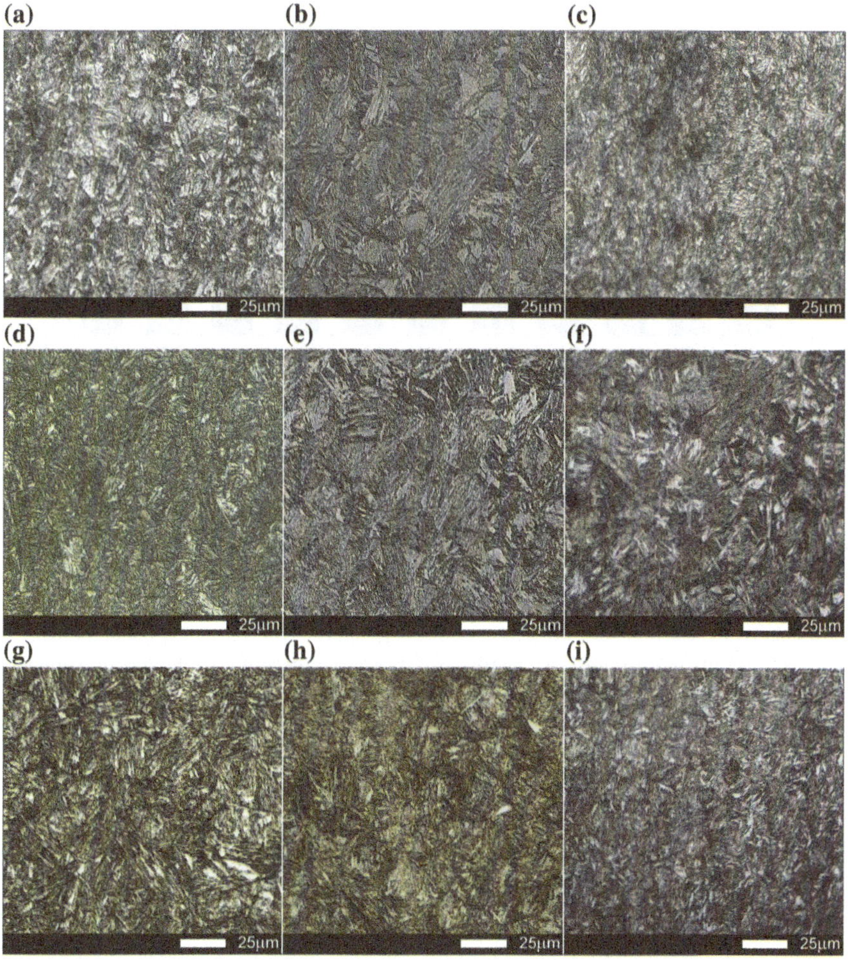

Fig. 17 Microstructures of low-alloy martensitic steels [12, 31, 40–42]: **a** Hardox 400—temper sorbite with areas of tempered martensite; **b** Brinar 400—hardening martensite with lath morphology; **c** HTK 900H—structure with post-martensitic morphology—temper sorbite with areas of tempered martensite; **d** Hardox 500—tempered martensite; **e** Brinar 500—fine-lath hardening martensite; **f** B27—tempered martensite, structure with post-martensitic morphology; **g** Hardox 600—hardening martensite with areas of tempered martensite; **h** Hardox Extreme—fine-lath hardening martensite with areas of fine-acicular martensite; **i** XAR 600—fine-lath hardening martensite with uniformly arranged areas of tempered martensite. Etched with 2% HNO_3, light microscopy

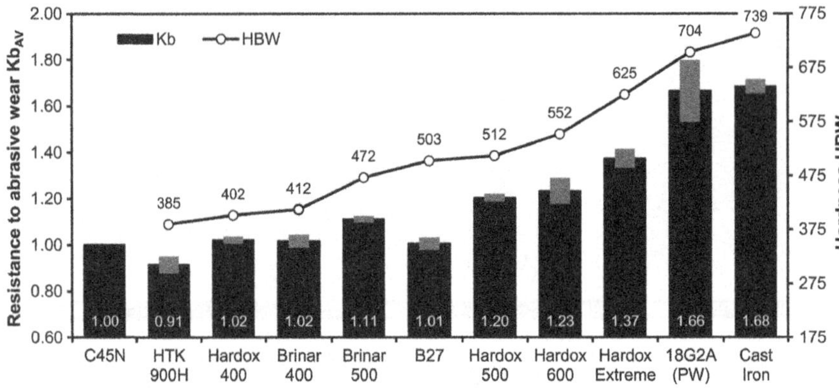

Fig. 18 Relative abrasion resistance coefficient and hardness measurements of selected low-alloy martensitic steels [31, 32, 43]: C45N—steel in normalised condition, hardness 220 HBW; 18G2A (PW)—steels with pad-welded layer of Fe–Cr–C alloy; Cast Iron—high-chromium cast iron

3.2 Abrasive Wear Resistance

With regard to the fact that abrasive wear resistance is the most exposed feature of the above-mentioned materials, results of laboratory testing of selected martensitic steels for friction with loose abrasive material are shown in Fig. 18.

The abrasive wear tests were carried out on a laboratory tester T-07 with use of aloxite 90. The results indicate that the materials with post-martensitic structures and hardness above 500 HBW (Fig. 18)—in the applied testing conditions—are similar. It is also worth to note that the obtained abrasive wear coefficients do not deviate significantly from those obtained for normalised unalloyed steel 45 with a much lower hardness of 220 HBW. This results from different mechanisms of abrasive wear of hard structures with post-martensitic orientation and two-phase ferritic-pearlitic structures characteristic for the steel C45N. This statement can be a recommendation to verify laboratory tests by field tests with consideration of changing load, in addition to abrasion processes.

However, in comparison to steel C45, clear increase of abrasion resistance can be observed for metallic materials with hardness much above 500 HBW. It is worth to note in this case, that values of relative abrasion resistance coefficient show very high correlation with material hardness. Such correlation is not observed for lower hardness levels, in spite of similar morphologies of these steels.

3.3 Analysis of Welded Joints

Besides resistance to abrasive wear, a feature very often emphasised by manufacturers of a group of steels is also a possibility to join them by welding. For the steels Hardox

400 and Hardox 500, most often used in conditions of abrasive wear, this parameter was repeatedly verified. For example, their good weldability was confirmed in [11] and [14], as well as possibility to obtain relatively high mechanical parameters of the performed welded joints.

However, it was demonstrated in the above-mentioned research that degradation of these steels takes place after welding, in heat-affected zones. The consequences are variable changes of hardness and local loss of abrasion resistance. It should be additionally stressed that the above-described thermal phenomena very often accompany other technological operations connected with processing and forming parts made of these steels. In this connection, the authors of this study have drawn the conclusion that, from practical point of view, it is worth to consider the issues of preparation and heat treatment of welded joints of both Hardox 400 and Hardox 500 that are recognised as steels with good weldability, as well Hardox 600 that exceeds the weldability criterion.

After analysis of the data given in Table 1, it can be generally said that Hardox steels 400, 450 and 500 (and also HTK 900H) are the materials with relatively good weldability, which is confirmed by their location in the C-CEV diagram (Fig. 19) close to the zone of low susceptibility to cracking (I) or susceptibility dependent on welding conditions (II). The other considered grades of Hardox steels, with real CEV equivalents much higher than 0.45, do not confirm this statement. The data concerning weldability of Hardox 550, Hardox 600 and Hardox Extreme steels are completed with very negative experiences of local distributors and fabricators of these steels. They unambiguously indicate a very limited or completely missing weldability of these steels, which results mostly from their susceptibility to brittle

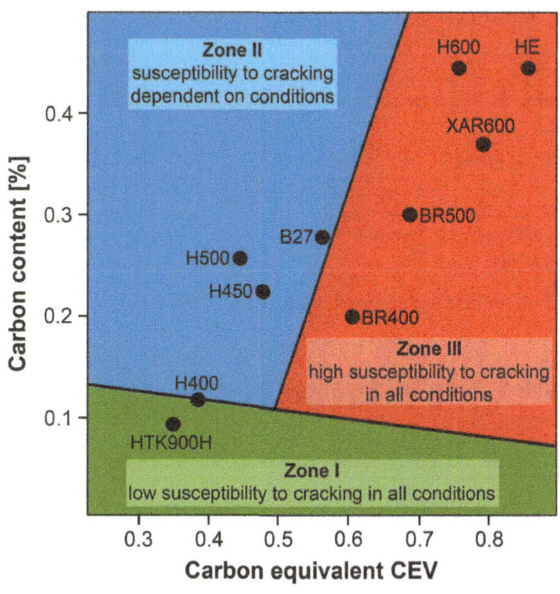

Fig. 19 Susceptibility of selected martensitic steels to cracking: H—Hardox, BR—Brinar. Own study based on data from Table 1 and from [44]

cracking during welding and from occurrence of wide zone with lower hardness that practically changes definition of these steels.

In the further part of this study, results of examinations concerning preparation and evaluation of welded joints of Hardox steels are presented. Because of a big variety of properties, the results are divided into two groups: Hardox 400 and 500, and separately Hardox 600. In both cases, properties of joints were analysed in as-welded condition (AW) and after heat treatment (HT). Basic properties of the used welding materials are given in Tables 3 and 4.

Welded joints of Hardox 400 and Hardox 500 were made with the arc submerged under flux (SAW), using welding consumables and parameters recommended by the manufacturer. In both cases, one kind of welding wire Multimet IMT9 ϕ 3 mm and flux Lincoln Electric FX780 were applied (Table 3). The joints were made on sheets 8 mm thick joined with double single-bead welds, without chamfering, at the following parameters [11, 14]:

- minimum current strength for individual welds: $I1 = 300$ A, $I2 = 500$ A;
- voltage of electric arc for both welds: $U = 30$ V;
- constant welding speed: $v = 0.35$ m/min;
- heat input: $Q = 2.74$ kJ/mm.

Welded joints of Hardox 600 were made by the TIG method using welding consumables and parameters recommended by the manufacturer (Table 4). Sheets 12 mm thick were joined with double multi-layer welds, using different weld metals for backing welds (A in Fig. 20b) and filling welds (B in Fig. 20b) at the following parameters [12]:

- non-consumable electrode: thoriated tungsten electrode;
- maximum current strength for individual weld layers: $I = 90$ A;

Table 3 Properties of welding consumables used for welding steels Hardox 400 and 500 [11, 14, 45]

Weld metal	C	Mn	Si	Cr	Ni	Mo	R_e	R_m	A_5	KCV_{-20}
	Chemical composition [%]						[MPa]		[%]	[J/cm^2]
IMT9 + FX780	0.09	–	–	–	–	–	403[a]	506[a]	31[a]	126[a]

[a]Data for flux with properties close to FX780

Table 4 Properties of welding consumables used for welding steel Hardox 600 [12]

Weld metal	C	Mn	Si	Cr	Ni	Mo	R_e	R_m	A_5	KCV_{-40}
	Chemical composition [%]						[MPa]		[%]	[J/cm^2]
OK Aristorod 89 (GMn4Ni2CrMo)	0.10	1.9	0.80	0.30	2.10	0.65	920	1000	18	60
UltraMag (G3Si1)	0.07	1.45	0.85	0.04	0.01	0.01	471	580	25	73

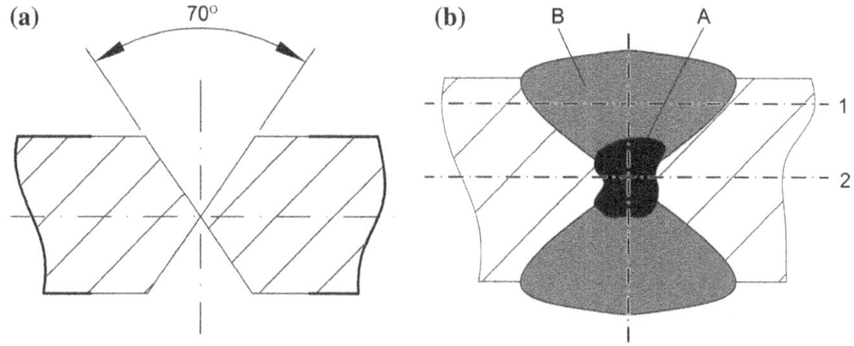

Fig. 20 Layout of welded joints of Hardox 600 [12]: **a** preparation of sheet edges; **b** application of weld metal: A—UltraMag, B—Aristorod 89; 1 and 2—lines of hardness measurements

- voltage of electric arc: $U = 9.5$ V;
- approximate welding speed: $v = 140$ mm/min;
- shielding gas: 99.99% argon;
- shielding gas flow: 10 l/min;
- interpass temperature: Ti < 400 °C;
- heat input: $Q < 0.5$ kJ/mm.

Table 5 presents schedules and parameters of carried-out thermal treatment of welded joints of Hardox steels and obtained mechanical properties. For Hardox 400 and Hardox 500, mechanical testing after additional operations of heat treatment was not carried out. Directly after welding, welded joints of these steels were characterised by medium tensile strength of 615 and 663 MPa, respectively. In relation to the strength of base materials, it is only ca. 50% and 43% of R_m for the steels Hardox 400 (1250 MPa) and Hardox 500 (1550 MPa). It should be emphasised that scatter of individual results was small, which confirms correctness of welded joints (absence of welding imperfections) and indicates, in the authors' opinion, improper selection criteria of materials and welding conditions.

Specimens of welded Hardox 600 in as-welded condition showed medium tensile strength of 871 MPa. Like the grades Hardox 400 and 500, this value makes only ca. 44% of strength of base material Hardox 600 (2000 MPa). However, the results obtained for this steel after heat treatment according to the schedules given in Table 5 need a more comprehensive comment. After heat treatment, tensile strength is higher, with an average of 1386 MPa. This makes ca. 70% of the value for Hardox 600 sheet in as-delivered condition. However, considering single values for individual specimens, it can be assumed that the upper limit for the applied technology is higher than 1500 MPa (specimen H600-HT-UTS-3 in Table 5), which makes at least 75% of nominal value for the base material. It is also worth to mention that, in the case of welding followed by heat treatment, highly favourable values of impact strength at ambient temperature were obtained for this material (42 J/cm^2), especially in the context of so high static tensile strength.

Table 5 Schedules, heat-treatment parameters and selected mechanical properties of welded joints of Hardox steels [12, 14]: AW—as-welded, HT—heat-treated, N—normalised, Q—quenched, T—tempered

Designation of specimen	Steel	Condition	Schedule and parameters of heat treatment	R_m [MPa]		KCV_{+20} [J/cm^2]
H400-AW-UTS-1	Hardox 400	AW	Untreated	618	615	–
H400-AW-UTS-2				620		
H400-AW-UTS-3				605		
H400-AW-UTS-4				620		
H400-AW-UTS-5				618		
H400-AW-UTS-6				608		
H500-AW-UTS-1	Hardox 500	AW	Untreated	669	663	–
H500-AW-UTS-2				668		
H500-AW-UTS-3				642		
H500-AW-UTS-4				668		
H500-AW-UTS-5				669		
H400-HT-HD H500-HT-HD	Hardox 400/500	HT	[Q + T] Q: 930°C/20 min/water T: 200°C/120 min/air			

(continued)

Table 5 (continued)

Designation of specimen	Steel	Condition	Schedule and parameters of heat treatment	R_m		KCV_{+20}	
				[MPa]		[J/cm^2]	
H600-AW-UTS-1	Hardox 600	AW	Untreated	917	871	–	
H600-AW-UTS-2				835			
H600-AW-UTS-3				862			
H600-AW-KCV-1				–		169.2	163
H600-AW-KCV-2						154.4	
H600-AW-KCV-3						164.6	
H600-HT-UTS-1	Hardox 600	HT	[N + Q + T] N: 900 °C/60 min/air Q: 930 °C/15 min/oil T: 200 °C/120 min/air	1266	1386	–	
H600-HT-UTS-2				1331			
H600-HT-UTS-3				1562			
H600-HT-KCV-1				–		47.8	42
H600-HT-KCV-2						39.7	
H600-HT-KCV-3						39.2	

Figures 21, 22, 23, 24, 25, 26, 27, 28, 29, 30, 31 show macroscopic images, hardness measurements and microstructures in characteristic zones of welded joints of steels Hardox 400, 500 and 600. It can be said on the grounds of the carried-our examinations that welding induces variable structural changes, manifested mostly by very wide zones with lower hardness (Fig. 23). In as-welded Hardox 400, this zone is ca. 70 mm wide and for Hardox 500 and 600 it reaches over 80 mm. In the case of welded joints of Hardox 400 and Hardox 500 width of this zone is decided mostly by the applied method of SAW welding but, in the case of Hardox 600 welded by TIG method, it is affected mostly by parameters and conditions of welding (e.g. interpass temperature). So, in the authors' opinion, weldability analysis of steels Hardox 400 (CEV = 0.38), Hardox 500 (CEV = 0.44) and especially Hardox 600 (CEV = 0.76) regarded as non-weldable indicates that welding technology of these very high-strength steels should obligatorily include post-process heat-treatment operations.

Analysis of available examination results [11] and [14] concerning welded joints of Hardox 400 and Hardox 500 showed that both steels are characterised with good

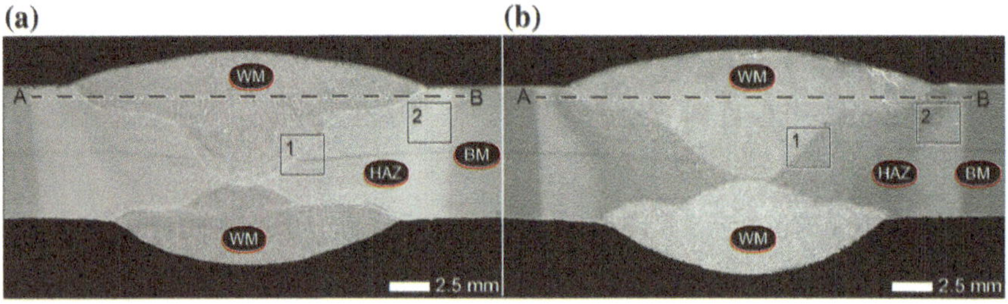

Fig. 21 Macroscopic views of welded joints of Hardox steel as-welded [11]: **a** Hardox 400; **b** Hardox 500; WM—weld material, HAZ—heat-affected zone, BM—base material; A–B—hardness measurement lines. Etched with 5% HNO_3, stereoscopic microscopy

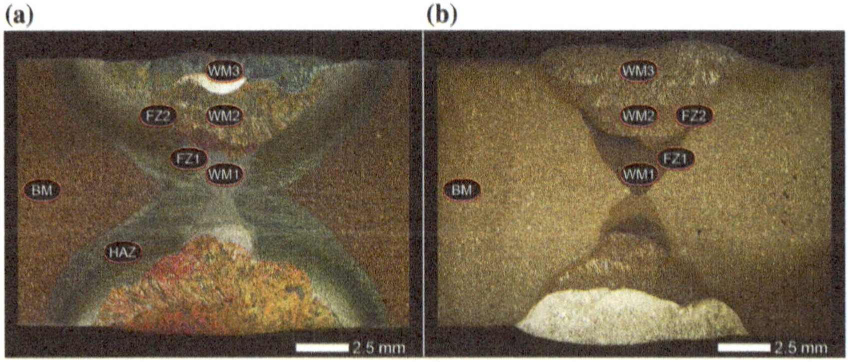

Fig. 22 Macroscopic views of welded joints of Hardox 600: **a** as-welded; **b** heat = treated; WM1—UltraMag, WM2/WM3—OK Aristorod 89, FZ—fusion zone, HAZ—heat-affected zone, BM—base material. Etched with 2% HNO_3 and electrolytically with chromium acid; light microscopy

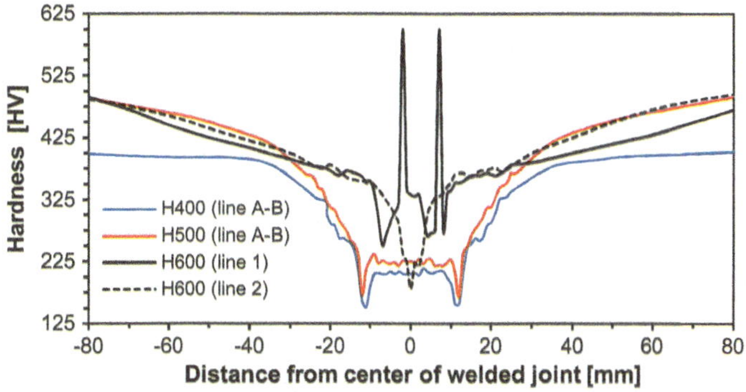

Fig. 23 Hardness distribution of Hardox welded joints as-welded along the lines 1 and 2 shown in Fig. 20 and along the line A-B shown in Fig. 21 [11, 12]

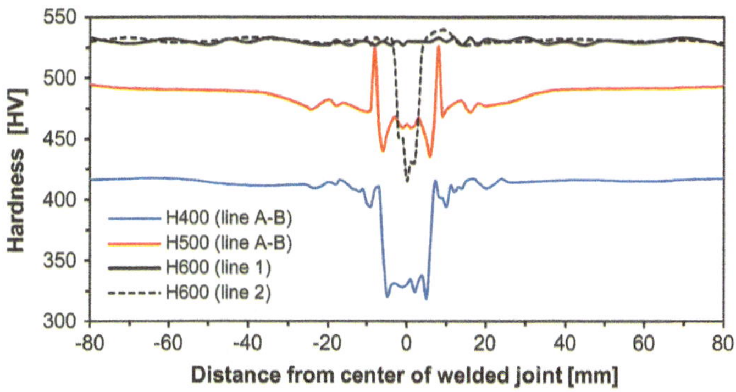

Fig. 24 Hardness distribution of Hardox welded joints after heat treatment along the lines 1 and 2 shown in Fig. 20 and along the line A–B shown in Fig. 21 [11, 12]

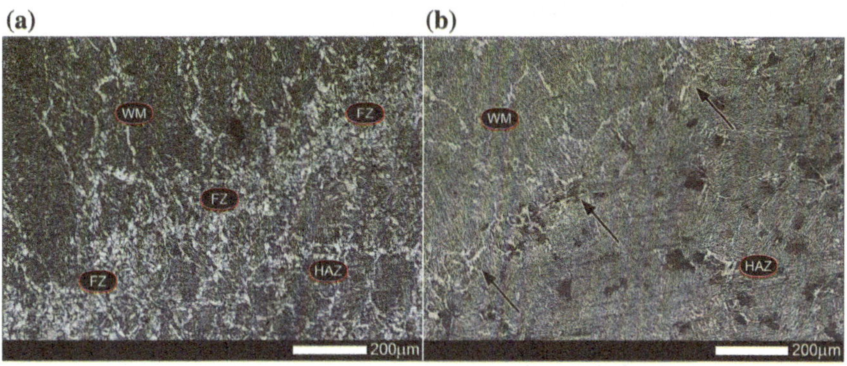

Fig. 25 Microstructures of Hardox welded joints in fusion zone (frame 1 in Fig. 21) [11]: **a** Hardox 400 as-welded—clearly outlined fusion line; **b** Hardox 500 as-welded—weakly outlined fusion line (marked with arrows); WM—weld material, FZ—fusion zone, HAZ—heat-affected zone. Etched with 2% HNO_3, light microscopy

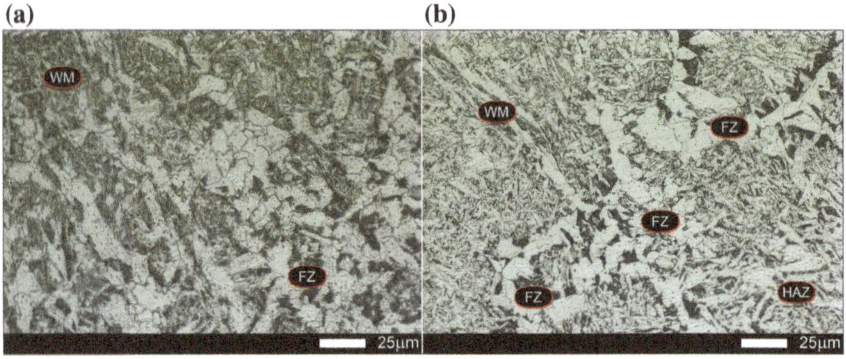

Fig. 26 Magnified microstructures of Hardox welded joints shown in Fig. 25 [11]: **a** Hardox 400 as-welded—granular and acicular ferrite with areas of quasi-eutectoid and bainite. Numerous carbides visible inside ferrite grains. **b** Hardox 500 as-welded—granular and acicular ferrite with few areas of quasi-eutectoid and carbide particles; WM—weld material, FZ—fusion zone, HAZ—heat-affected zone. Etched with 2% HNO_3, light microscopy

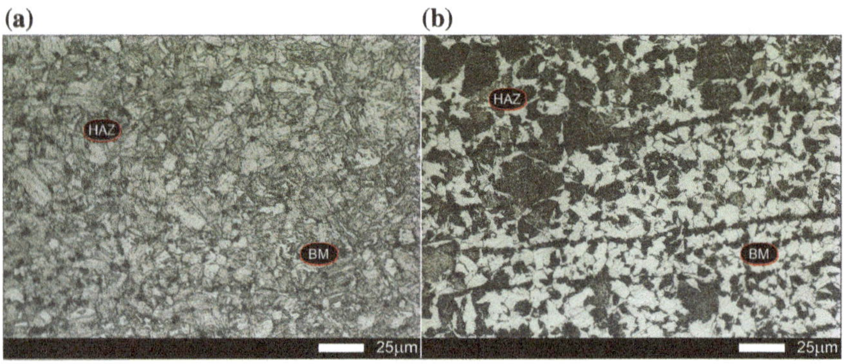

Fig. 27 Microstructures of Hardox welded joints in heat-affected zone (frame 2 in Fig. 21) [11]: **a** Hardox 400 as-welded—tempered martensite similar to microstructure of Hardox 400 sheet as-delivered; **b** Hardox 500 as-welded—granular pearlite with lamellar cementite. In the zone next to base material, visible is structure similar to normalised structure—ferritic-pearlitic with band-like arranged pearlite; HAZ—heat-affected zone, BM—base material. Etched with 2% HNO_3, light microscopy

weldability. In both considered cases, weld seams were characterised by correct depth of fusion and missing welding defects affecting their mechanical properties. Welded joints made of Hardox 400 were characterised by correct (typical for this grade) hardness distribution (Fig. 23) and clearly outlined fusion zone revealed by light microscopy (Fig. 25a). In fusion zone, the weld showed structure of granular and acicular ferrite (Fig. 26a). Inside ferrite grains, numerous carbide particles were observed. In addition, colonies of quasi-eutectoid and areas of bainite were observed

(a) (b)

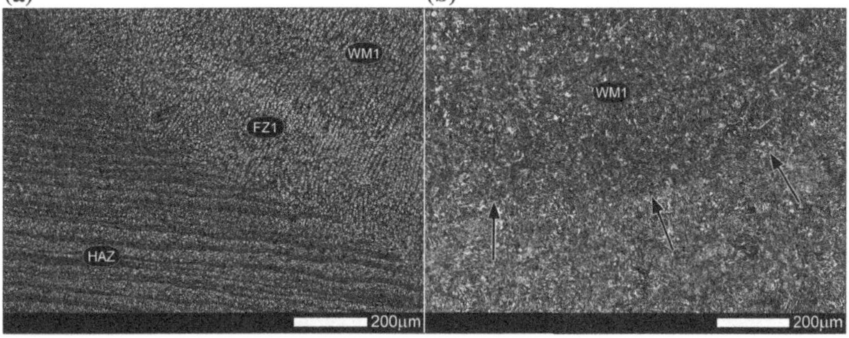

Fig. 28 Microstructures of Hardox 600 welded joints in fusion zones marked as FZ1 in Fig. 22 [12]: **a** as-delivered—wide, clearly outlined fusion zone; **b** heat-treated—weakly outlined fusion line (marked with arrows); WM1—UltraMag, FZ—fusion zone, HAZ—heat-affected zone. Etched with 2% HNO_3, light microscopy

(a) (b)

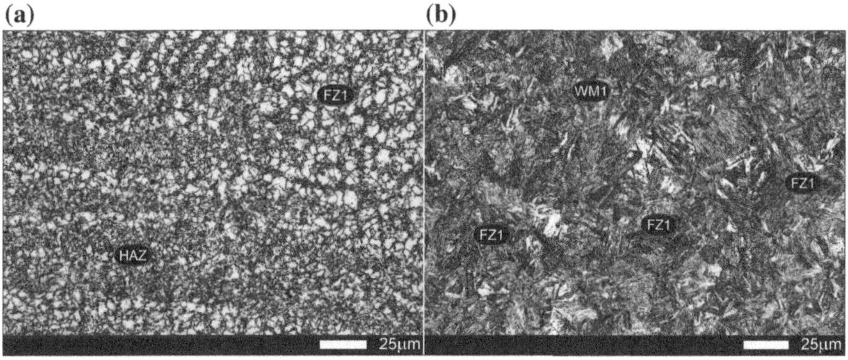

Fig. 29 Magnified microstructures of Hardox 600 welded joints shown in Fig. 28 [12]: **a** as-delivered—ferritic-pearlitic-martensitic structure diversified within the specimen; **b** heat-treated—martensitic structure with diversified morphology; WM1—UltraMag, FZ—fusion zone, HAZ—heat-affected zone. Etched with 2% HNO_3, light microscopy

in this zone. However, refinement of structure, characteristic for normalisation processes, was observed in heat-affected zone (Fig. 27a).

In the case of welded joints of Hardox 500, good weldability was also confirmed, although deviating from that of Hardox 400. In the condition directly after welding, the weld was characterised by hardness distribution on cross-section typical for welded martensitic steels (Fig. 23), as well as by fusion line much narrower and less outlined than in Hardox 400 (Fig. 25b). In this zone, welded joint of Hardox 500 showed structure of granular and acicular ferrite (Fig. 26b). A small number of carbide particles was also observed inside ferrite grains. In addition, colonies of quasi-eutectoid, troostite and bainite could be seen in selected areas of fusion zone.

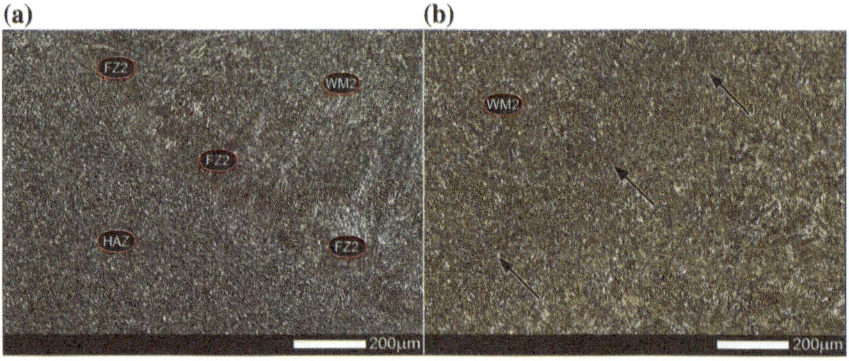

Fig. 30 Microstructures of Hardox 600 welded joints in fusion zones marked as FZ2 in Fig. 22 [12]: **a** as-delivered—clearly outlined fusion zone; **b** heat-treated—structurally homogenised fusion zone (weakly outlined fusion line marked with arrows); WM2—OK Aristorod 89, FZ—fusion zone, HAZ—heat-affected zone. Etched with 2% HNO_3, light microscopy

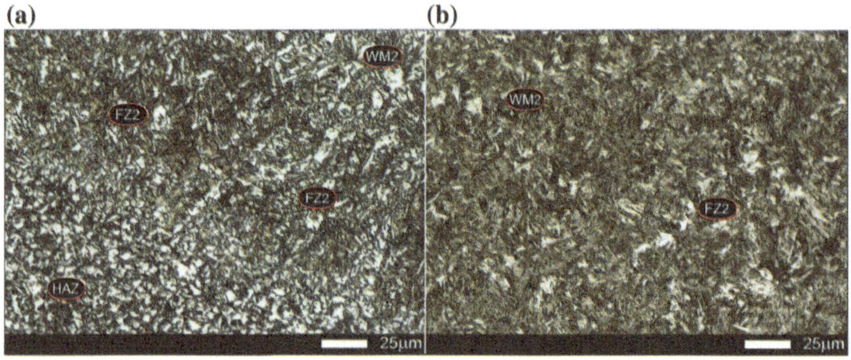

Fig. 31 Magnified microstructures of Hardox 600 welded joints shown in Fig. 30 [12]: **a** as-delivered—in fusion zones and in base material—ferritic-martensitic structure with small areas of bainite. In HAZ—ferritic-pearlitic structure; **b** heat-treated—martensitic structure with small areas of bainite; WM2—OK Aristorod 89, FZ—fusion zone, HAZ—heat-affected zone. Etched with 2% HNO_3, light microscopy

Heat-affected zone of Hardox 500 welded joint was characterised by ferritic-pearlitic structure with granular and laminar pearlite, see Fig. 27b.

Structural changes caused by thermal welding processes in welded joints of Hardox 600 were found much more differentiated than those observed in Hardox 400 and 500, with respect to both type and morphology of microstructure [12]. In the zone of weld metal UltraMag (zone WM1 in Fig. 22), microstructure is composed of granular ferrite with areas of granular pearlite (Fig. 28a). As a result of imposition of subsequent layers, this zone changes from fine-grained normalised structure in the middle of the weld, through equilibrium ferritic-pearlitic zone to dendritic structure resulting from directional crystallisation. Morphology of structure in this zone made it possible to obtain hardness of 180 HV only (Fig. 23). Heat treatment resulted in significant structural homogenisation of this zone: the obtained structure consisted of tempered martensite with areas of bainite (Figs. 28b and 29b). Analogous considerations related to the zone of weld metal OK Aristorod 89 (WM2/WM3 in Fig. 22) lead to the conclusion that, in as-welded condition, microstructure of this zone consists of tempered martensite with acicular grains of ferrite and sparse colonies of troostite. As a result of post-welding heat treatment, the structures composed almost completely of tempered martensite with bainitic areas (Figs. 30b and 31b). Structural changes in the welded joint of Hardox 600 resulted in uniform hardness distribution on the entire cross-section of the weld (line 1 in Fig. 24). Relatively small scatter of hardness (144 HV) was recorded in middle part of heat-treated weld—in weld metal UltraMag of low hardness (WM1 in Fig. 22) where hardness values ranged between 392 and 536 HV. In as-welded condition, scatter of hardness was 306 HV (between 157 and 463 HV).

In the case of welded joints of Hardox 600, it is worth to pay attention to structural changes in fusion zones. After welding, a very clearly outlined fusion line can be distinguished (Figs. 25a and 26a) in the zone FZ1 (marked in Fig. 22). Observed are also other structural changes, adversely affecting mechanical properties: two structures of various morphologies are directly adjoining. One of them is band-like ferritic-pearlitic-martensitic structure of base material, and the other is dendritic structure of weld metal (Figs. 28a and 29a). It is worth to emphasise that the revealed banding of base material in heat-affected zone does not result from manufacturing process of Hardox 600 sheet, but is caused exclusively by welding processes as a result of diffusion of chemical components. Additional heat-treatment operations carried-out after welding made it possible to homogenise the entire fusion zone FZ1 to fine-lath tempered martensite with bainitic areas (Figs. 28b and 29b). Structural changes similar to these above-described are observed also in fusion zone with the weld metal OK Aristorod 89 (FZ2/FZ3 in Fig. 22). In as-welded condition, fusion zones FZ2 and FZ3 on their complete length are composed of martensitic structure with small part of bainite and acicular ferrite (Figs. 30a and 31a). After heat treatment, like in previous cases, these zones became homogenised to the structure of lath martensite (Figs. 30b and 31b).

In connection with the before-mentioned results of laboratory testing of abrasive wear resistance of selected metallic materials, it is worth to cite results of service tests of basic brown-coal mining machines in real service conditions. Therefore, basic assumptions and results of field tests carried out in the mine BOT KWB Turów S. A. with use of fixed chute of the excavator KWK-1500s (K-9) are presented below. For the tests, the following assumptions were accepted [14]:

- In the field tests, lining plates of Hardox 400 (2 pcs. 400 × 200 × 12) and Hardox 500 (2 pcs. 400 × 200 × 15) were used;
- Properties of the tested constructional materials were previously recognised in laboratory examinations;
- Hardox plates were arranged together with the surrounding lining made of pad-welded steel 18G2A;

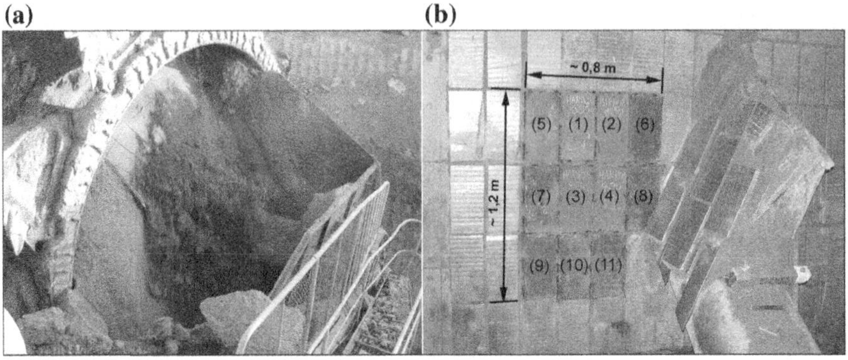

Fig. 32 Chute of bucket wheel of excavator KWK-1500s (K-9) [14]: **a** general view; **b** assembly diagram of lining plates. Materials of plates: 1, 2—Hardox 500; 3,4—Hardox 400; 5–11—pad-welded steel 18G2A

- Location of Hardox plates shall consider concurrent abrasive and dynamic action of the excavated material on lining plates;
- All the testes lining plates shall be situated in the area with identical intensity of influence of the excavated material;
- Lining plates shall be fitted with the same technology, i.e. with positional welds;
- It must be possible to monitor the tested plates for the entire time of the field experiment.

Figure 32 shows a fragment of left trapezoidal chute of bucket wheel together with assembly diagram of experimental plates of Hardox 400 and Hardox 500. The plates were fitted in the same way as that used for fitting pad-welded plates 18G2A, i.e. with positional welds. Figures 33, 34, 35, 36 show photographic documentation from final phase of the experiment, i.e. 565 h. At that stage, analysis of wear of individual plates could be only performed for plates Hardox 400 and pad-welded plates 18G2A. However, degree of wear of lining plates made of Hardox 500 was so low that these plates could not be subjected to replacement or evaluation.

Evaluation of durability of individual lining plates indicates, like it was stated in [14], that plates of Hardox 400 (Nos. 3 and 4 in Fig. 32b) were abraded evenly on their entire working planes, and no defects like pits, losses or cracks appeared. However, pad-welded plates (Nos. 10 and 11 in Fig. 32b) were subjected to uneven wear, which resulted in numerous macroscopic defects on their surfaces. Results of the whole experiment of wear of lining plates are presented in a comparative way in Table 6.

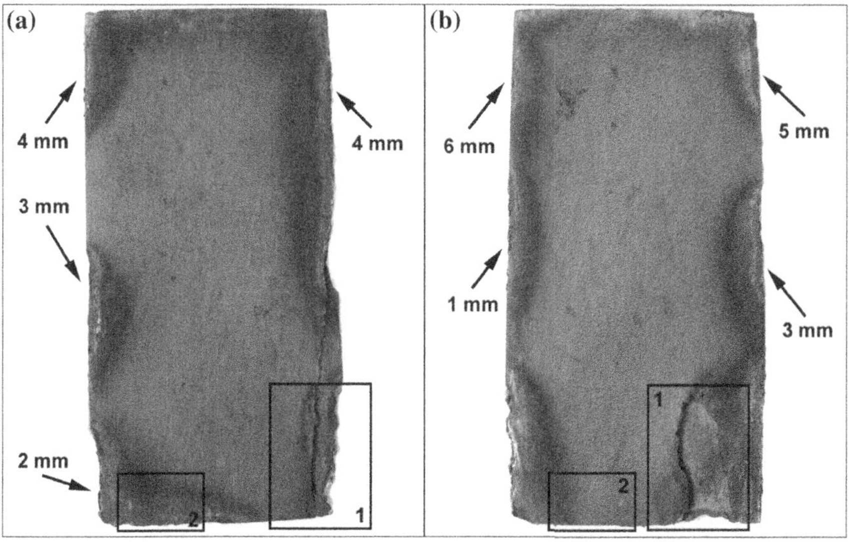

Fig. 33 Macroscopic view of surfaces of Hardox 400 plates after 595 h of experiment [14]: **a** plate No. 3; **b** plate No. 4. Uniform nature of wear with no identified macroscopic defects

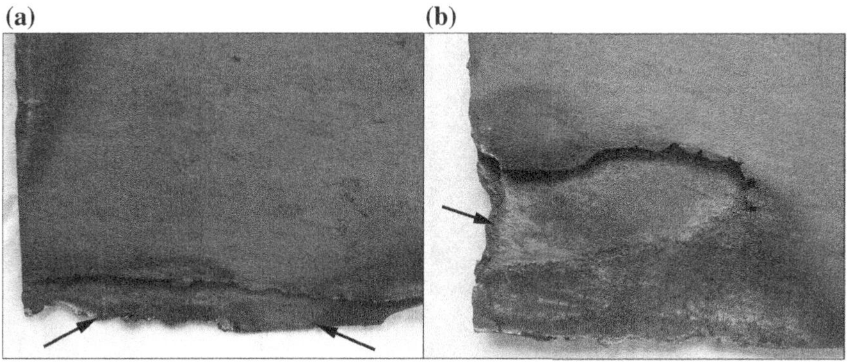

Fig. 34 Magnified fragments of plates marked with frames 1 in Fig. 33 [14]: **a** fragment of plate No. 3; **b** fragment of plate No. 4. Arrows indicate substrate material uncovered as a result of complete wear of plate

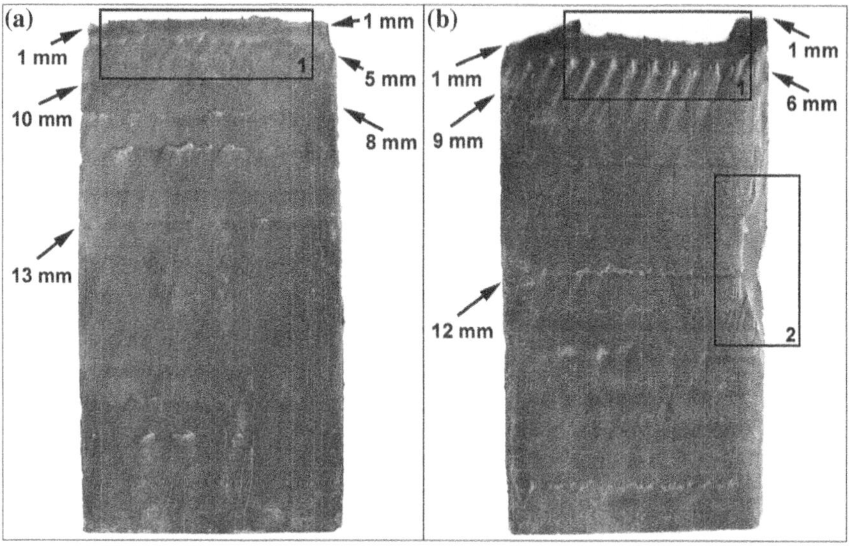

Fig. 35 Macroscopic view of surfaces of pad-welded plates of 18G2A after 595 h of experiment [14]: **a** plate No. 10; **b** plate No. 11. Uneven nature of wear with numerous defects in form of macrocracks, furrows and spallings of padding weld

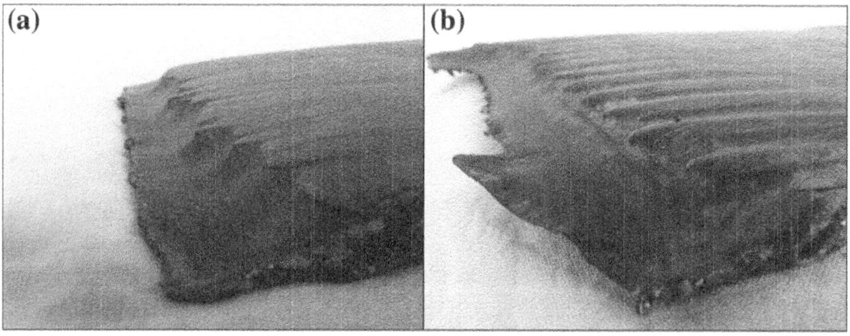

Fig. 36 Magnified fragments of plates marked with frames 1 in Fig. 35 [14]: **a** fragment of plate No. 10; **b** fragment of plate No. 11. Deep abrasion on entire cross-sections of plates together with substrate material is visible

Table 6 Course of wear of lining plates on fixed chute of the excavator KWK-1500s (K-9) [14]

Work time [h]	Material of lining plates		
	Pad-welded steel 18G2A	Hardox 400	Hardox 500
0	Corrugated surface of padding weld, seams transverse to direction of material flow, single macrocracks	Uniform surface with no damages, covered with thin layer of atmospheric corrosion products	Uniform surface with no damages, covered with thin layer of atmospheric corrosion products
85	No cracks were observed, condition similar to that observed before	Uniform bright surface with no damages	Uniform bright surface with no damages
180	Developing cracks of padding weld at the edges, single zones of complete wear of padding weld material	Light rounding of corners in places of joining with substrate	Uniform bright surface with no damages
255	Progressive wear of padding weld, worn-away seams, extended zones with no padding weld	Developed zones with increased wear of plates at the edges	Uniform bright surface with no damages
460	No padding weld at the plate edge 200 mm long, rounding of edges, development of macrocracks	Uniform wear of lining, thickness reduced from 15 to 8 mm, no damages	Uniform wear of lining, thickness reduced from 15 to 13 mm, no damages
531	Further uncovering of substrate material, new cracks appear along seams of welding pad, extractions of padding weld on seams	Uncovered surface of substrate on right lower corner of the plate No. 3, remaining area with no damages	Uniform wear of lining, thickness reduced from 15 to 13 mm, no damages
565	Further uncovering and abrasion of substrate material, cracks along seams of welding pad, extractions of padding weld on seams	Increase of uncovered substrate surface area by 100% in relation to the previous condition, remaining area with no damages	Uniform wear of lining, thickness reduced from 15 to 12 mm at the edge, no damages
595	Complete wear of plates No. 10 and 11, replacement required	Dismantling of plates No. 3 and 4, no traces of wear of substrate material	Uniform wear of lining, thickness reduced from 15 to 12 mm at the edge, no damages
1416	Plates installed in place of Hardox 400 linings qualify for replacement	Dismantled and replaced with pad- welded plates	Uniform bright surface with no damages

4 Summary

The above-presented deliberations on mechanisms of degradation processes and wear of constructional materials in selected assemblies of mining machines lead to the conclusion that low-alloy martensitic steels compose a modern group of constructional materials showing much better service properties than those offered by the steels so-far applied for parts of basic mining machines.

In the authors' opinion, these steels can make an alternative for pad-welded low-alloy steels, high-chromium cast iron or surface-hardening steels that, in hitherto prevailing solutions, are used for linings of chutes and transfer stations, as well as for parts of buckets of brown-coal excavators.

However, with respect to this material group, a very important issue are diverse structural changes occurring in the material during welding. As a result of these changes, the steels lose—in welded areas—resistance to abrasive wear that is additionally intensified by dynamically changing loads.

Therefore, application and complete use of high durability indices of these steels require implementing a specialised laboratorial strategy to the processes of design and manufacture of selected parts of mining machines. This strategy mostly concerns application of welding techniques for joining sheets of the considered group of steels and should ensure (reproduce) their high resistance to abrasive wear processes.

Thus, the issues related to welding of low-alloy martensitic steels, defined also as abrasion-resistant steels, have been discussed extensively in the presented study.

References

1. Dombrowskij, N.G.: Excavators. Izd. Maszinostrojenie, Moscow (1961)
2. Hawrylak, H., Sobolski, R.: Basic Machines of Surface Mining. Publishing House "Śląsk", Katowice, 1970 (in Polish)
3. Rasper, L., Bahr, J., Felkel, E.: Gartner, E. Der Schaufelradbagger als Gewinungsgerat. Trans. Tech. Publications (1973)
4. Dudek, D., Oziemski, S., Sobczykiewicz, W.: Elements of degradation theory of machines. Problemy Maszyn Roboczych, No. 4, Warszawa (1994) (in Polish)
5. Dudek, D., et al.: Maintenance Strategy of Surface-Mining Machines and Facilities with High Level of Technical Degradation. Publishing Office of Wrocław University of Science and Technology (2013) (in Polish)
6. Pękalski, G.: Materials aspects of the degradation theory conception of a multi-criterion system of estimation of the state of material. Syst. J. Transdisciplinary Syst. Sci. **3**(2), 108–129 (1998)
7. Lesiuk, G., Szata, M.: Aspects of structural degradation in steels of old bridges by means of fatigue crack propagation. Mater. Sci. **47**(1), 82–88 (2011)
8. Pękalski, G.: Causes and results of corrosion of brown-coal excavators. Górnictwo Odkrywkowe (Surface Mining) (5–6), 72–76 (2002) (in Polish)
9. Konat, Ł., Pękalski, G.: Structures, properties and applications of high-strength wear-resisting martensitic steels. Visnik Chmel'nickogo Nacional'nogo Universitetu **1**(4), 196–205 (2007)
10. Kowalczyk, M.: Dimensioning of welded supporting structures of basic surface-mining machines with respect to fatigue life. Ph.D. Thesis (22.06.2010), Wrocław University of Science and Technology, Institute of Machinery Design and Operation, Wrocław (2010) (in Polish)

11. Frydman, S., Konat, Ł., Pękalski, G.: Structure and hardness changes in welded joints of Hardox steels. Arch. Civ. Mech. Eng. 2008, 8, 15–27
12. Konat, Ł., Białobrzeska, B., Białek, P.: Effect of welding process on microstructural and mechanical characteristics of Hardox 600 steel. Metals **7**(9), 1–18 (2017)
13. Pękalski, G.: Adapted to users' needs concept of evaluation of degradation of steel. Conference materials: Problems and innovations in energetic overhauls—PIRE 2000. III Scientific-technical Conference, Szklarska Poręba, 28 XI–1 XII, Wrocław, pp 169–177 (2000) (in Polish)
14. Konat, Ł.: Structures and properties of Hardox steels and their application possibilities in conditions of abrasive wear and dynamic loads. Ph.D. Thesis, Wroclaw University of Technology, Wroclaw, Poland (2007) (in Polish)
15. Leda, H.: Structural Aspects of Mechanical Properties of Selected Materials. Publishing Office of Poznań University of Technology, Poznań (1998). (in Polish)
16. Pękalski, G.: Corrosion of body structure of basic machines of surface mining. Problemy Maszyn Roboczych **7**(z. 7), 75–89 (1996) (in Polish)
17. Kwiatkowska-Szygulska, B., Mikołajczyk, A., Ostrycharz, D., Dziewanowski, M., Hanula, P.: Condition of environment in the trangle borderland on the grounds of a research within National Environment Monitoring. PECKIANA (9), 1–20 (2014) (in Polish)
18. Głuszko, M., Konat, Ł., Pękalski, G., Rabiega, J.: Corrosion and structural degradation processes in steel used for construction of Bridges. Fiziko-Himičeskaâ Mehanika Materialov spec. vyp. (4), 181–186 (2004)
19. Michońska, K.: Metallic materials used for a component of Central Railway Station in Wrocław. Scope and cost of overhaul. Master's Thesis (tutor G. Pękalski) (2013) (in Polish)
20. Haimann, R., Krajczyk, A., Pękalska, L., Pękalski, G.: Analysis of mechanical changes and development of technology of optimum annealing of welded parts (caterpillar girders, crawler treads, rocker arms etc.). Reports Inst. Mater. WUST, Ser. SPR No. 6 (1986) (in Polish)
21. Haimann, R., Krajczyk, A., Pękalska, L., Pękalski, G.: Influence of stress relieving on structure and properties of joints of cast steel with steels 18G2A and 12G2ANB. Weld. Technol. Rev. **8**, 7–8 (1990). (in Polish)
22. Krajczyk, A., Pękalska, L., Pękalski, G.: Problems of selection and heat treatment of material for track links of brown-coal mining machines. Surf. Min., R. 33, No. 1, 30–39 (1991) (in Polish)
23. Pękalski, G.: Forming structures and properties of castings of manganese-silicon cast steel by heat treatment. Acta Metallurgica Slovaca **2**, 214–219 (2001). (in Polish)
24. Dudek, D., Frydman, S., Huss, W.M., Pękalski, G.: The L35GSM cast steel—possibilities of structure and properties shaping at the example of crawler links. Arch. Civ. Mech. Eng. **11**(1), 19–32 (2011)
25. Krajczyk, A., Pękalska, L., Pękalski, G.: Problems of selection of materials with high abrasion resistance on the example of bucket teeth of excavators. Surf. Min., R. 33, No. 1, 40–47 (1991) (in Polish)
26. Krajczyk, A., Pękalska, L., Pękalski, G.: Optimisation of microstructures of materials used for bucket teeth of excavators. Experimental methods in manufacture and operation of machinery and means of transport. In: Scientific conference, Wrocław-Szklarska Poręba, 24–28 May 1993, Part 2, Wrocław: Publishing Office of WUST, 135–140 (1993) (in Polish)
27. Pękalski, G., Dudziński, W., Sachadel, U., Alenowicz, J.: Some macro- and microstructural aspects of pad welded layer durability. Arch. Civ. Mech. Eng. **5**(2), 85–104 (2005)
28. Cegiel, L., Kozerska, A., Pękalski, G.: Structure and resistance to abrasive wear of materials used for lining plates of chutes and dumpers in brown coal excavators. Surf. Min. **48**(5–6), 56–62 (2006). (in Polish)
29. Napiórkowski, J., Pękalski, G., Kołakowski, K.: Examination of structures and wear of pad-welded coatings in soil abrasive mass. Tribology R. 43, (3), 111–117 (2012) (in Polish)
30. Napiórkowski, J., Konat, Ł., Pękalski, G., Kołakowski, K.: Structures of the EL-Hard padding welds as a function of the abrasive wear resistance. Arch. Foundry Eng. **13**(1), 73–82 (2013)
31. Capanidis, D., Konat, Ł., Pękalski, G., Sachadel, U., Wieleba, W.: Possibilities of using Hardox steels, their structures and selected properties. In: XVIII Scientific conference, Zakopane, 17–20

January 2005, Kraków: Publishing Office of Cracow University of Technology, pp 59–69 (2005) (in Polish)
32. Łętkowska, B.: Influence of heat treatment on structure and selected properties of B27 and 28MCB5 Steels. Ph.D. Thesis, Wroclaw University of Technology, Wroclaw, Poland (2013) (in Polish)
33. Hardox—Das Verschleißblech der vielen Möglichkeiten; Information materials of SSAB Oxelösund (2002)
34. Hardox wear plate-Wear and abrasion resistant steel–SSAB. Available online: https://www.ssab.com/products/steel-categories/wear-resistant-steels (accessed on 9 Mar 2017)
35. Ilsenburger Grobblech GmbH—Material Specification Sheets. Available online: http://www.ilsenburger-grobblech.de/en/products/material-specification-sheets.html (accessed on 18 Sept 2017)
36. The Steel Service Centre–Flinkenberg Steel. Available online: http://www.flinkenberg.fi/wp-content/uploads/DATASHEET-XAR600.pdf (accessed on 18 Sept 2017)
37. Hardox wear plate, abrasion resistant steel-Stal-Hurt. Available online: http://www.stal-hurt.com (accessed on 18 Sept 2017)
38. HTK Katowice—abrasion resistant steel plate. Available online: http://www.htk.com (accessed on 20 Sept 2017)
39. Ruukki—Hot rolled hardenable boron steels. Available online: http://www.ruukki.com (accessed on 25 Apr 2016)
40. Konat, Ł., Napiórkowski, J., Białobrzeska, B.: Structural properties and abrasive-wear resistance of Brinar 400 and Brinar 500 stels. Tribologia **48**(3), 67–75 (2017)
41. Konat, Ł., Napiórkowski, J., Kołakowski, K.: Resistance to wear as a function of the microstructure and selected mechanical properties of microalloyed steel with boron. Tribologia **47**(4), 101–114 (2016)
42. Napiórkowski, J., Lemecha, M., Konat, Ł.: Effect of external load on the process of steel consumption in a soil mass. Tribologia **48**(3), 111–117 (2017)
43. Białobrzeska, B., Konat, Ł.: Comparative analysis of abrasive-wear resistance of Brinar and Hardox steels. Tribologia **48**(2), 7–16 (2017)
44. Graville, B.A.: Cold Cracking in welds in HSLA steels, welding of HSLA (Microalloyed) structural steels. In: Proceedings of the AIM/ASM Conference, Rome, Italy, 9–12 Nov (1976)
45. Multimet—welding consumables. Available online: http://www.multimet.com.pl/ (accessed on 26 Sept 2017)

Modelling, Computing, and Analyzing Large-Size Rotary Joints

Tadeusz Smolnicki

Abstract The chapter presents selected issues regarding modelling, calculating, and researching large-size rotary joints utilized in heavy equipment, vehicles, and engineering objects. It is the fruit of 30 years of cooperation with the industry and research conducted during these years, associated with designing new machines, modernizing machines after long service, diagnosing bearings' technical state, and forecasting their remaining time of operation. Ways of modelling the support subassembly-slewing bearing-support subassembly system with the use of finite elements are presented. These models were utilized to identify the distribution of load carried by rolling elements and to evaluate the impact of that distribution on the stiffness of the support subassembly. Based on the obtained results, a method of bearing correction greatly reducing the effort of a bearing was developed. It was applied to two heavy equipment machines. Original measuring systems built for that purpose confirmed the correctness of the method. Numerical models used for forecasting plastic wear of a bearing's raceway and researching this phenomenon in surface mining machines are also described.

Keywords Rolling bearings · Slewing bearings · Load distribution · Bearing correction · Plastic wear

1 Introduction

The application of bearings in order to move various large objects has been a vital need since antiquity, but at the same time, it has been a difficult problem and a real challenge for designers, manufacturers, and operators. These objects include heavy equipment, military vehicles, and engineering objects, such as bridges, etc.

Slewing bearings of all sorts are the current trend in the realization of rotational motion. Such bearings transmit all three components of forces and tilting moments around any radius of the bearing (Fig. 1).

T. Smolnicki (✉)
Faculty of Mechanical Engineering, Wroclaw University of Science and Technology, ul. Wybrzeze Wyspianskiego 27, 50-370 Wroclaw, Poland
e-mail: tadeusz.smolnicki@pwr.edu.pl

M. Sokolski (ed.), *Mining Machines and Earth-Moving Equipment*,
https://doi.org/10.1007/978-3-030-25478-0_3

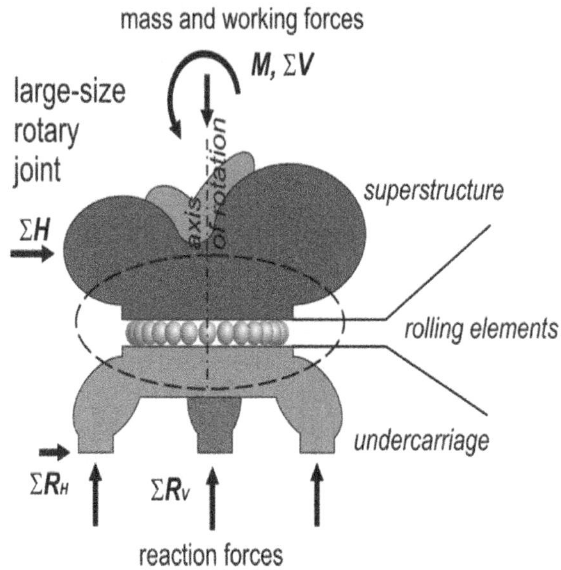

Fig. 1 Large-size rotary joint

Such bearings are often equipped with a toothed-wheel rim, which enables rotational motion of one element relative to the position of the other. An example of such a technical solution is shown in Fig. 2.

The Faculty of Mechanical Engineering at Wrocław University of Science and Technology has actively participated in the development of theory, computational methods, engineering design processes, and technical condition evaluation in terms

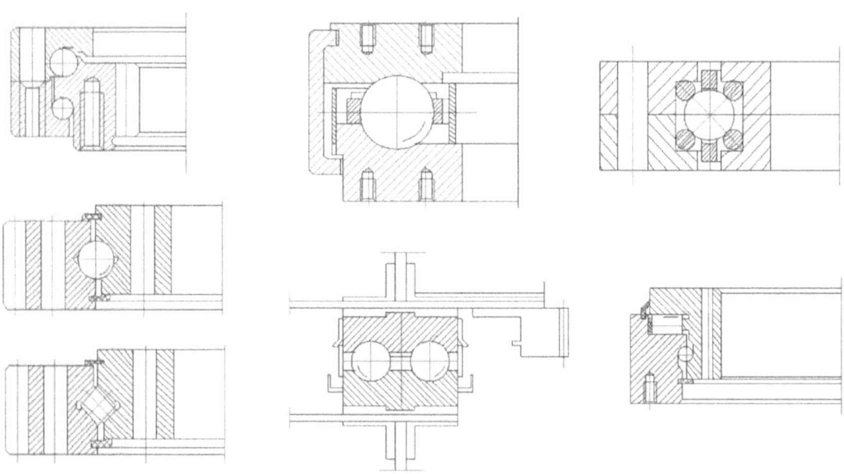

Fig. 2 Various structural modules of slewing bearings

of such bearings. The first works were realized by Nogieć and Malcher [1], Nogieć and Smolnicki [2].

The most fundamental problems in the application of this type of bearings are the distribution of forces transmitted onto individual rolling elements of the bearing and the impact that the stiffness of supporting constructions and the distribution of that stiffness have on the distribution of forces. These bearings contain from a few dozen to several hundreds of rolling elements which makes them statically indeterminate.

An example of an undercarriage portal frame's structural module is shown in Fig. 3. This structure is very heterogeneous—it is stiffer around the supports and less stiff between them.

Figure 4 shows the comparison between the stiffness of a single rolling element and the stiffness of a support subassembly. The comparison was carried out based on an analysis of several dozen existing bearings exploited in bucket wheel excavators, spreaders, stacker-reclaimers, and crawler excavators. If the pitch diameter of a bearing exceeds 3 m, the impact of the support subassembly's stiffness must be taken into account.

Problems with the exploitation of surface mining machines utilized in Polish mines drove the evolution of computational methods and research methods, as well

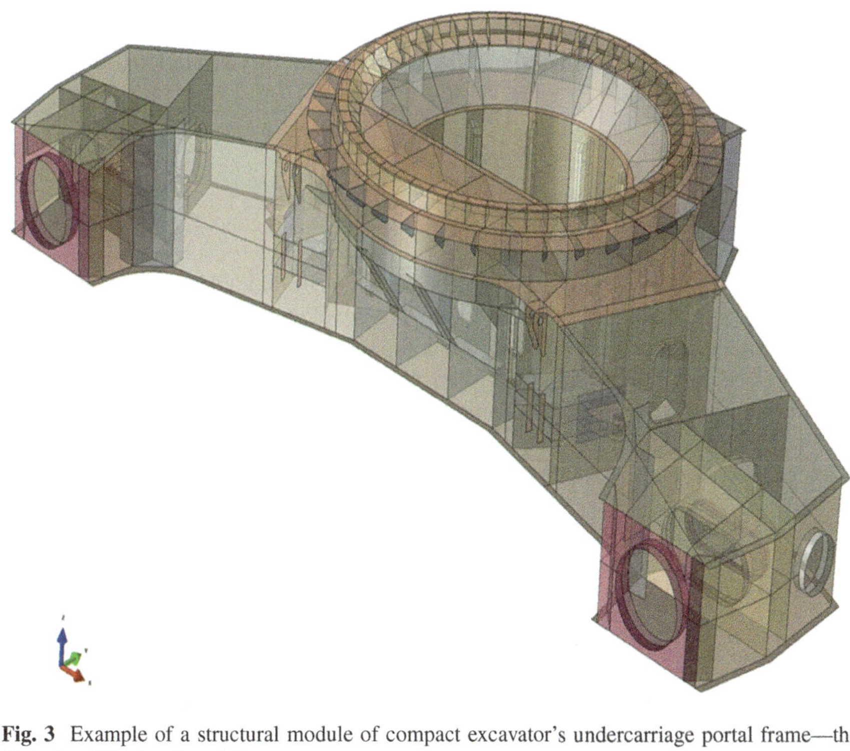

Fig. 3 Example of a structural module of compact excavator's undercarriage portal frame—the diameter of the bearing: 4.8 m

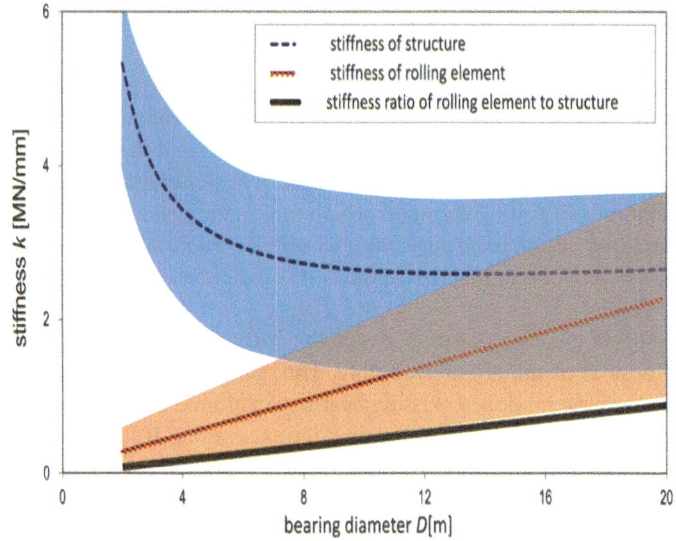

Fig. 4 The comparison between the stiffness of a single rolling element and the stiffness of structures

as imposed the development of design criteria and their in situ verification in operating conditions.

2 Computational Methods

Computational methods used for slewing bearings can be divided into two categories: classical and numerical. Classical methods are developed based on static equilibrium equations for the raceways of both components, which take into account the force-displacement characteristics of the rolling element-raceway system and omit the deformability of support subassemblies, which in case of heavy equipment is a name that is most commonly used for describing the machine's superstructure and undercarriage. Numerical methods utilize the principles of Finite Element Method.

2.1 Classical Methods

Due to the limited computational capabilities of classical methods, the models of slewings have to be simplified. The bearing is what remains of the original model, while the structures of support subassemblies and the flexural deformability of the bearing ring are omitted. Load distribution of such models is also simplified. Ohnrich's method and Matthias's method, which have been most commonly used, also omit many factors, such as the interdependence of impacts of individual loads carried by the bearing (axial forces, yaw moments, and radial loads) on the forces transmitted by individual rolling elements and the change in their angles of operation.

Because of the present ability to solve complex systems of equations numerically, a method of evaluating the internal forces in a multi-row ball-bearing loaded with force V acting on the eccentricity e in the axial direction and loaded with radial force H, the vector of which does not have to be located in the plane of operation of the tilting moment caused by force V. The bearing can be equipped with any number of rows of balls and can have either internal clearance or clamp in the axial and radial direction. The symbols in the equations were given according to Fig. 5. The top ball's centre is located in point C:

$$C\left(R\ 0\ h\right) \tag{1}$$

The placement of centres of curvature of the raceways A and B:

$$\begin{aligned} &A\left(R + (r_A - r)\sin\gamma\ 0\ h + (r_A - r)\cos\gamma\right), \\ &B\left(R - (r_B - r)\sin\gamma\ 0\ h - (r_B - r)\cos\gamma\right). \end{aligned} \tag{2}$$

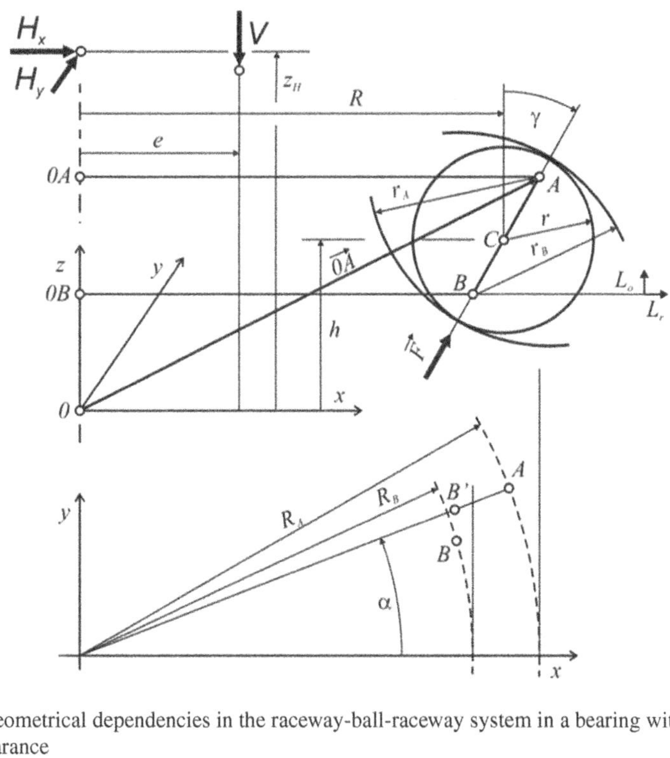

Fig. 5 Geometrical dependencies in the raceway-ball-raceway system in a bearing with axial and radial clearance

Having regarded the axial and radial clearance:

$$\begin{aligned} &A\big(R + (r_A - r)\sin\gamma \;\; 0 \;\; h + (r_A - r)\cos\gamma\big), \\ &B_L\big(R - (r_B - r)\sin\gamma + L_r \;\; 0 \;\; h - (r_B - r)\cos\gamma + L_o\big). \end{aligned} \tag{3}$$

The placement of rolling elements on raceway A results from the pitch of the bearing; the i-th point's coordinates are:

$$A\Big((R + (r_A - r)\sin\gamma)\cos\big(2\pi \tfrac{i-1}{z}\big) \;\; (R + (r_A - r)\sin\gamma)\sin\big(2\pi \tfrac{i-1}{z}\big) \;\; h + (r_A - r)\cos\gamma \Big). \tag{4}$$

where i is the index of a rolling element and z is the quantity of rolling elements.

The coordinates of a point on raceway B corresponding to the i-th point are:

$$B_L\Big((R + (r_A - r)\sin\gamma + L_r)\cos\big(2\pi \tfrac{i-1}{z}\big) \;\; (R + (r_A - r)\sin\gamma)\sin\big(2\pi \tfrac{i-1}{z}\big) \;\; h + (r_A - r)\cos\gamma + L_o \Big). \tag{5}$$

Ring A is treated as stationary, ring B has 5 degrees of freedom:

- axial displacement u_z
- horizontal displacement u_x and u_y
- rotation α_x and α_y around x and y axes.

Due to displacement, points B_i, which are related with the ring 0_B and correspond to the rolling elements of the ring 0_B, are moved. It is necessary to evaluate new points of intersection B'_i of the moved ring $0'_B$ with the half-plane Π_{Ai} designated by the axis A_i and the line segment $0A_i$.

The equation of the half-plane Π_{Ai} has the following form:

$$x\sin\left(2\pi\frac{i-1}{z}\right) - y\cos\left(2\pi\frac{i-1}{z}\right) = 0. \tag{6}$$

Having regarded the clearance, the coordinates of a point on the ring 0_B with an angular coordinate α are:

$$B_L\Big((R + (r_A - r)\sin\gamma + L_r)\cos(\alpha) \;\; (R + (r_A - r)\sin\gamma)\sin(\alpha) \;\; h + (r_A - r)\cos\gamma + L_o \Big).$$

The new placement of point B, after rotating it by α_x and α_y and translating it by a vector $\{u_x, u_y, u_z\}$ can be written in the following form:

$$B' = \begin{Bmatrix} x' \\ y' \\ z' \end{Bmatrix} = \begin{bmatrix} \cos\alpha_y & 0 & \sin\alpha_y \\ 0 & 1 & 0 \\ -\sin\alpha_y & 0 & \cos\alpha_y \end{bmatrix} \begin{bmatrix} 1 & 0 & 0 \\ 0 & \cos\alpha_x & -\sin\alpha_x \\ 0 & \sin\alpha_x & \cos\alpha_x \end{bmatrix} \begin{Bmatrix} (R + (r_A - r)\sin\gamma + L_r)\cos(\alpha) \\ (R + (r_A - r)\sin\gamma)\sin(\alpha) \\ h + (r_A - r)\cos\gamma + L_o \end{Bmatrix} + \begin{Bmatrix} u_x \\ u_y \\ u_z \end{Bmatrix} \tag{7}$$

For every i-th rolling element, the following should be found iteratively:

$$\alpha_i : x'(\alpha_i)\sin\left(2\pi\frac{i-1}{z}\right) - y'(\alpha_i)\cos\left(2\pi\frac{i-1}{z}\right) = 0, \tag{8}$$

next, coordinates of the point should be evaluated:

$$B_i' = \begin{Bmatrix} x_i' \\ y_i' \\ z_i' \end{Bmatrix} \tag{9}$$

The change in the distance between raceways equals:

$$\delta_i = \left|A_i B_i'\right| - \left|A_i B_i\right| \tag{10}$$

With the use of the value δ_i, the force F_i transmitted by a rolling element can be evaluated:

$$F_i = \begin{cases} \left(\frac{\delta_i}{C}\right)^{\frac{1}{\eta}} E d^{2-\frac{1}{\eta}} & \delta w > 0 \\ 0 & \delta \le 0 \end{cases} \tag{11}$$

where C and η are dimensionless constants of the raceway-rolling element-raceway system's force-displacement characteristic, and d is the diameter of a rolling element.

The vector of the force transmitted by a rolling element can be written as:

$$\vec{F_i} = F_i \frac{\overrightarrow{B_i' A_i}}{\left|B_i' A_i\right|} \tag{12}$$

If the ring B is loaded with axial force V, which is placed on eccentricity e, and horizontal force H, composed of H_x and H_y components, applied at height z_H, then the conditions for equilibrium can be written down:

$$\sum X = 0 \sum Y = 0 \sum Z = 0 \sum M_x = 0 \sum M_y = 0 \tag{13}$$

The sum of the rolling elements' forces acting upon the ring of the bearing can be written as:

$$\begin{Bmatrix} -V \\ H_x \\ H_y \end{Bmatrix} + \sum_{j=1}^{m}\sum_{i=1}^{z_j} \overrightarrow{F_{ji}} = 0 \tag{14}$$

while the sum of moments acting upon the ring B is:

$$\begin{Bmatrix} -H_y z_H \\ Ve + H_x z_H \\ 0 \end{Bmatrix} + \sum_{j=1}^{m}\sum_{i=1}^{z_j} \overrightarrow{0A} \times \overrightarrow{F_{ji}} = 0 \tag{15}$$

The system of equations shown above should be solved iteratively in order to find such generalized displacements of the ring *B*: u_x, u_y, u_z, α_x, α_y that the work of external forces *W* acting upon the bearing reaches the maximum:

$$\left(u_x, u_x, u_z, \alpha_x, \alpha_y\right) : W_V + W_{H_x} + W_{H_y} = \max \tag{16}$$

It is then possible to evaluate forces F_{ji} in individual rolling elements and the change in the angle of operation of individual rolling elements:

$$\Delta\gamma = \arccos \frac{AB' \cdot AB}{|AB'| \cdot |AB|} \tag{17}$$

This method can be utilized for calculating bearings with a maximum diameter of 3 m or as an initial calculation for bearings with bigger diameters. The method requires iterative solving.

2.2 Numerical Methods Based on FEM

The first bearing whose force distribution on rolling elements was evaluated with the use of FEM in 1998 was the bearing of a stacker that had a pitch diameter of 10 m and contained 189 balls with a diameter of 150 mm. The machine's superstructure weighed 970 Mg. A special substitute element was used for calculating the Eq. (11). The substitute element was described in detail in work [3]. Its schematic diagram is shown in Fig. 6. The use of an element with a nonlinear, asymmetric stiffness

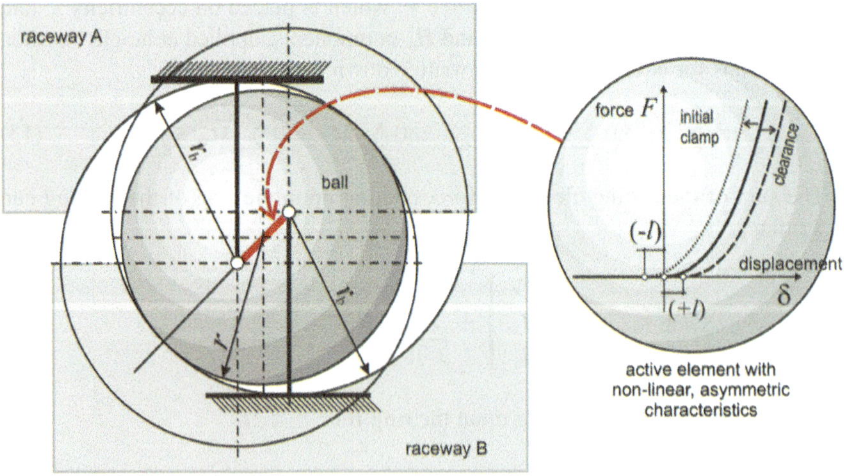

Fig. 6 Schematic diagram of a substitute element for modelling the raceway-ball-raceway system in slewing bearings

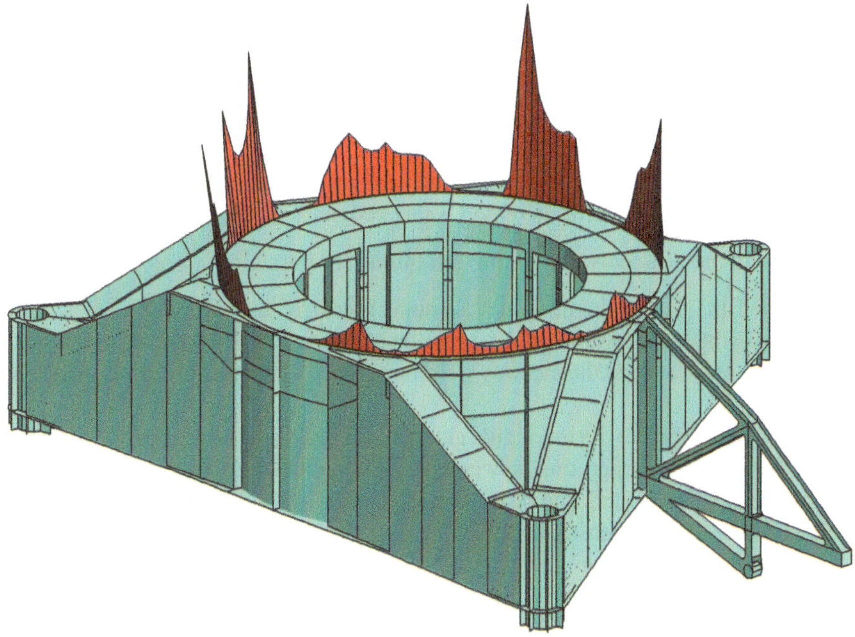

Fig. 7 Distribution of forces acting upon rolling elements of the stacker's bearing [4]

characteristic, capable of changing its direction relatively to the position of the ring of a bearing, allowed for a full recreation of the phenomena taking place in slewing bearings.

The distribution of load transmitted from the stacker's superstructure to individual rolling elements of the bearing was evaluated. The visualization of that distribution is shown in Fig. 7. The distribution was found to be highly irregular. The value of maximum load carried by a rolling element was three and a half times larger than the load evaluated with the use of classical methods with stiff support subassemblies factored in. The direction of operation of rolling elements deviated from the nominal direction by $\pm 15°$.

Discrete models of a bearing, which are made of the replacement models of the raceway-rolling element-raceway system and the discrete models of support subassemblies, represent the new quality in the methods of analyzing bearings. All of the significant phenomena, which take place both in macro-(analysis of the whole system) and micro-scale (analysis of a rolling element's and raceway's contact point), can be modelled. Discrete models of bearings developed with the use of substitute elements take into account the following factors:

- deformation of support subassemblies,
- heterogeneous input (output) of the load into (out of) the area of the bearing,
- any position of the structures over and under the bearing (superstructure and undercarriage),

- initial deplanation and ovalization of the bearing's rings caused by manufacturing and/or assembly errors,
- clearance (initial clamp) in the rolling element-raceway system,
- changes in the angle of operation of rolling elements,
- any force-displacement characteristic of the rolling element-raceway system,
- variable constructional (in places where segments are joined together) and exploitational (in places of increased wear) geometry of the raceway.

Thanks to these models, the following actions are possible:

- combined analysis of the machine's systems borne by the bearing,
- evaluating the load (effort) and the direction of operation for every rolling element and every place around the raceway's circumference for single- and multi-raceway bearings,
- analysis of the impact the individual geometrical parameters of support subassemblies have on the cooperation between the rolling elements and the raceway,
- correction of ring girders and bearing's raceways based not only on qualitative but also on quantitative measurements,
- simulation of the bearing's long-lasting work, evaluation of the wear's impact on load distribution.

The previously used models either did not allow for such advanced analysis or limited the range of the analysis.

The considerable advantage of the currently used models is the ability to utilize the same discrete models of support subassemblies which are used in FEM strength analysis in order to build the model of a bearing.

3 Example Results of Applying FEM

FEM models allow for the analysis to be carried out in any positions of the borne units relative to each other. Figure 8 presents an example of a load carried by rolling elements of a bearing of a bucket-wheel excavator in a selected position of the superstructure relative to the position of the undercarriage. The bearing has a diameter of 8 m and contains 105 balls with a diameter of 200 mm. A dominant impact of the global form of the superstructure's and undercarriage's structure can be seen, which determines the placement of the unloaded and overloaded zones of the bearing. The impact of the local supporting elements, such as ribs and diaphragms, is also visible. Every local stiffening causes an increase in the load carried by a rolling element, be it placed over or under it.

Through the analysis of results for the system in positions rotated by the angle equal to the pitching of the rolling elements of the bearing, an envelope, which is the distribution of maximum values of forces during one full revolution of the bearing, can be acquired (Fig. 9). The extreme values are evaluated in such a way. In zones of high stiffness, the load is over three times greater than in susceptible zones. Apart

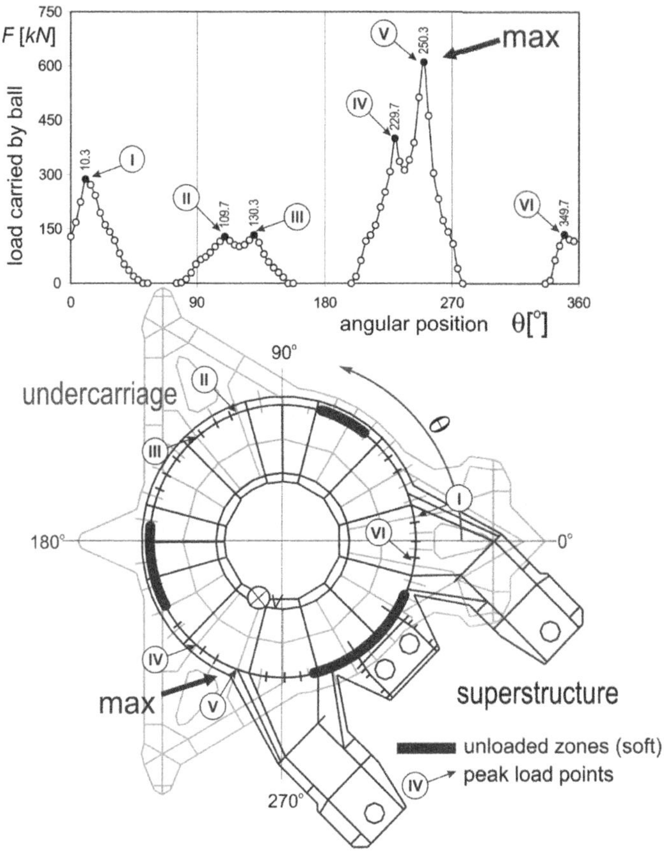

Fig. 8 Load carried by rolling elements of a bucket wheel excavator's bearing in selected positions

from the value of the forces, their direction is also evaluated. Figure 10 shows the vectors of forces acquired at 105 different positions of the superstructure relative to the position of the undercarriage in subsequent cross-sections of the undercarriage raceway. The length of line segments corresponds to the value of the force, while the direction of line segments corresponds to the angle of operation of a rolling element.

Slewings from a few dozen machines were analyzed in a similar manner. The consequence of support subassemblies' considerable susceptibility is a considerable overload of rolling elements when they are located between the zones of higher stiffness and underload when the balls happen to be located between these zones. In most of the analyzed bearings, a three- to a four-fold concentration of loads was acquired near the points of entry of supports. It is associated with the occurrence of higher structural stiffness in these places. Any local stiffenings were found to be disadvantageous for the bearing. They overlapped with the global distribution of stiffness and cumulated load in their ambiance.

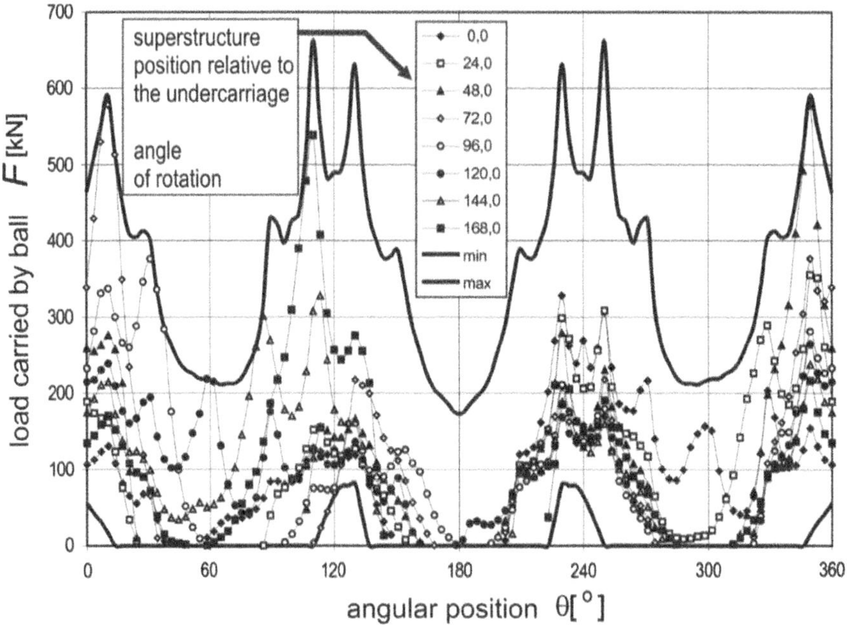

Fig. 9 The envelope of load distribution for different positions of the superstructure

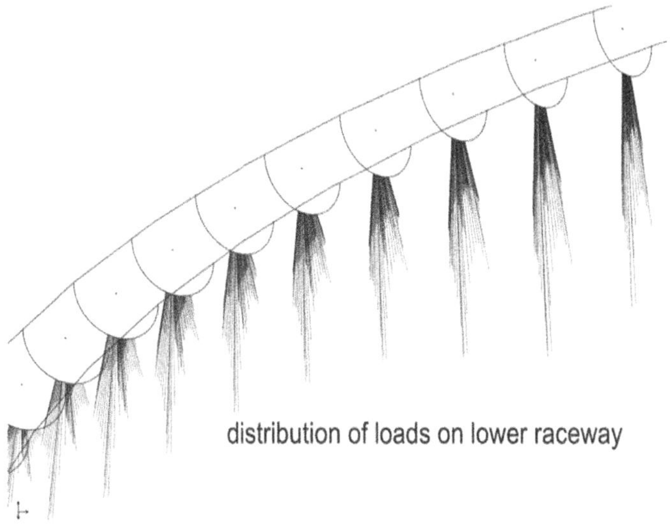

Fig. 10 Distribution of loads in subsequent cross-sections of the bearing's raceway

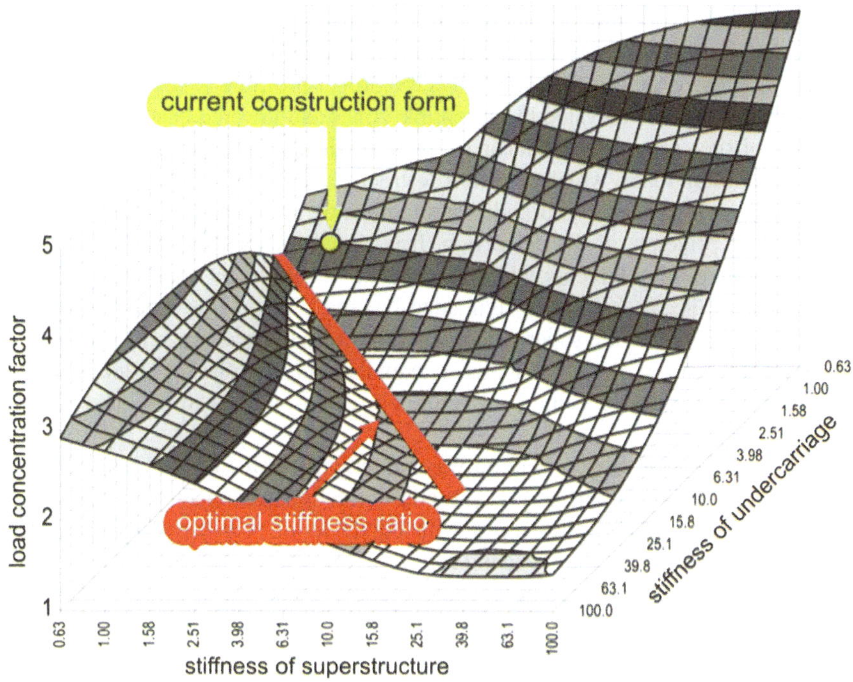

Fig. 11 A comparison of the impact of the stiffness of the superstructure and the undercarriage on the load concentration factor (red—optimal stiffness ratio)

A comparison of the most advantageous ratio of the stiffness of the superstructure to the stiffness of the undercarriage was conducted. The stiffness was being changed in a wide range and the value of load concentration was evaluated. The results are shown in Fig. 11. The area of the optimal ratio of the stiffness of the superstructure to the stiffness of the undercarriage was marked with the colour red. It was stated that in the analyzed case, the stiffness of the undercarriage portal frame had to be increased by 25% and the stiffness of superstructure platform had to be decreased by 33%. In the result, the maximum load carried by a rolling element dropped by 10%.

4 Correction of Raceway's Geometry

The analyses of multiple slewings showed that decreasing the load carried by the maximally efforted rolling elements in typical structures of support subassemblies would require that the mass of these elements be increased several times, which would be technically and economically indefensible, due to the exerted ground pressure and the required power of the installed drives. Decreasing the load carried by rolling elements was crucial because of the durability of the slewing.

In the year 2001, Przybyłek and Smolnicki proposed a new method for correcting raceway's geometry based on the introduction of an initial slope angle of the bearing's raceway [5]. The essence of the method is the proper profiling of the bottom of the raceway's groove. It is elevated in susceptible zones and lowered in stiffer zones. In the result of such procedure, load carried by individual rolling elements is equalled and the maximum loads, which are often destructive, are greatly reduced. Thanks to such treatment, the bearing's durability is significantly increased. Figure 12 presents the operating principle of the correction. It is aimed at equalling the load carried by individual rolling elements while the bearing is loaded by applying an initial deplanation to the raceway of an unloaded bearing. Due to the variable value of net load transmitted by the superstructure and the variable eccentricity of this force relative to the bearing's axis, the loads are equalled only for one set of loads, while the level of loads relative to the load before correction is reduced for other sets of load.

The correctness of this conception was numerically tested by Przybyłek [6] for three different machines. The ability to test the correction in a practical way appeared due to the low durability of the bearing with a diameter of 10 m used for rotating the superstructure of a stacker produced in 1999 with a mass of 970 Mg. After only 5 years of exploitation, the plastic wear of raceways accumulated to the value of 10 mm. The distribution of load carried by individual rolling elements of this machine as shown in Fig. 7. The raceways of such bearings are made from normalized steel, e.g. Ck45,

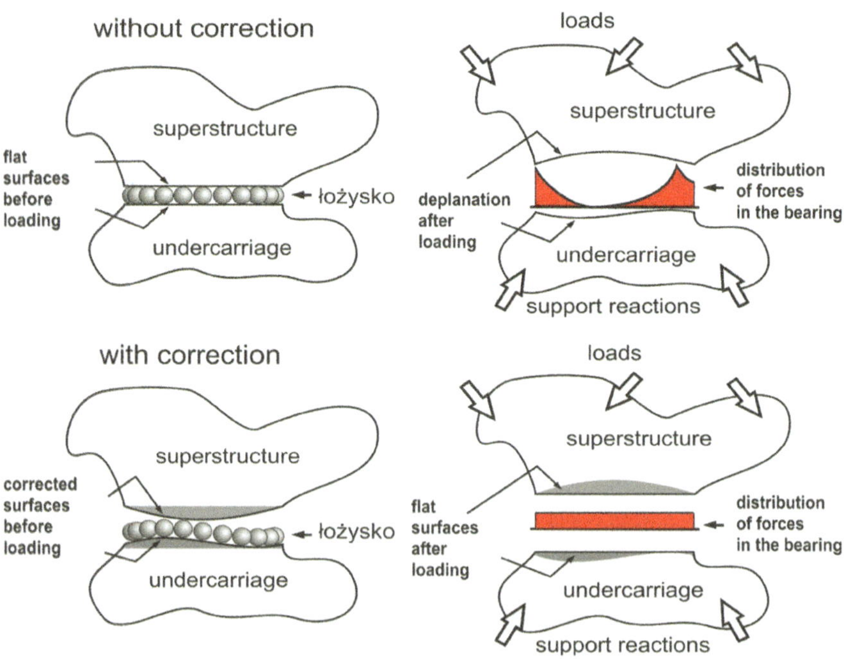

Fig. 12 The operating principle of geometrical correction of the bearing's raceway

Fig. 13 ZGOT 11500.100 Z-48 stacker at Turów surface mine

or from steel which was treated with heat to achieve smaller values of hardness, e.g. 42CrMo4V.

During the production of a new machine in 2007 (Fig. 13), a correction based on giving the raceway the inverse of the shape of the deflection curve of a support subassembly that was axially loaded with 120% of the superstructure's weight was proposed.

Figure 14 presents the geometrical model of the machine's slewing. Many numerical calculations with the use of FEM were conducted for both the corrected and not corrected raceway. The calculations were done for various eccentricities of load and different positions of the superstructure relative to the position of the undercarriage.

The comparison of the specific load carried by the bearing's balls along its circumference before and after the correction are shown in Fig. 15. The specific load is equal to force divided by the rolling element's diameter squared. It is a good measurement of the rolling element-raceway system's effort, independent from bearing's diameter.

As a result of applying the correction, the distribution was significantly evened. Under axial load, the balls in the zones of smaller stiffness are more efforted. It is connected to the fact the correction was carried out at 120% of the specific load.

The histogram of loads carried by balls changed after the correction (Fig. 16). The maximum specific load carried by a single ball decreased from 14.2 to 4.9 MPa. At the same time, the number of optimally (between 3 and 4 MPa) loaded balls increased.

The achieved level of loads guarantees a long-standing work of the slewing. The measurements of the geometry of the rings after the superstructure was built and the machine was ballasted showed that while the machine was loaded, it moved closer to the plane than when it was unloaded. 13 strain gauges were built into the bearing.

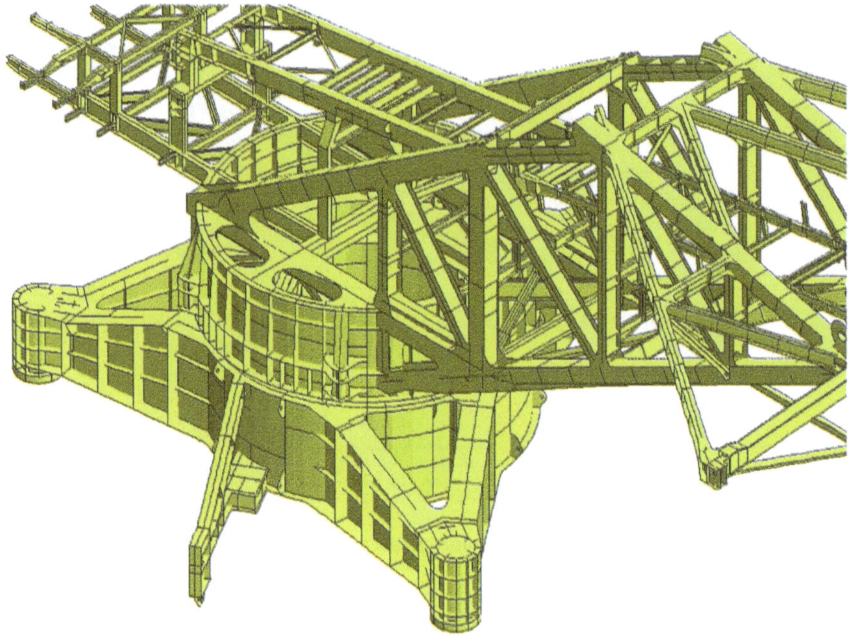

Fig. 14 Geometrical model of the ZGOT 11500.100 stacker's slewing

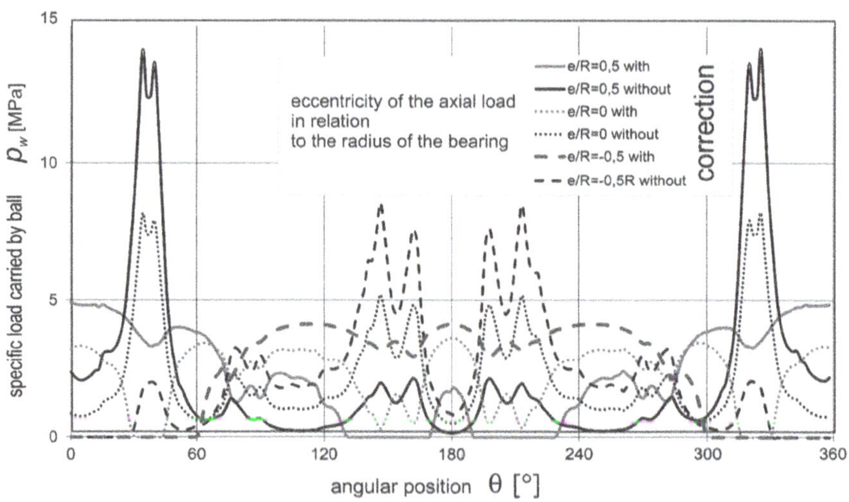

Fig. 15 The distribution of load (specific load) carried by balls of the stacker's bearing with and without correction

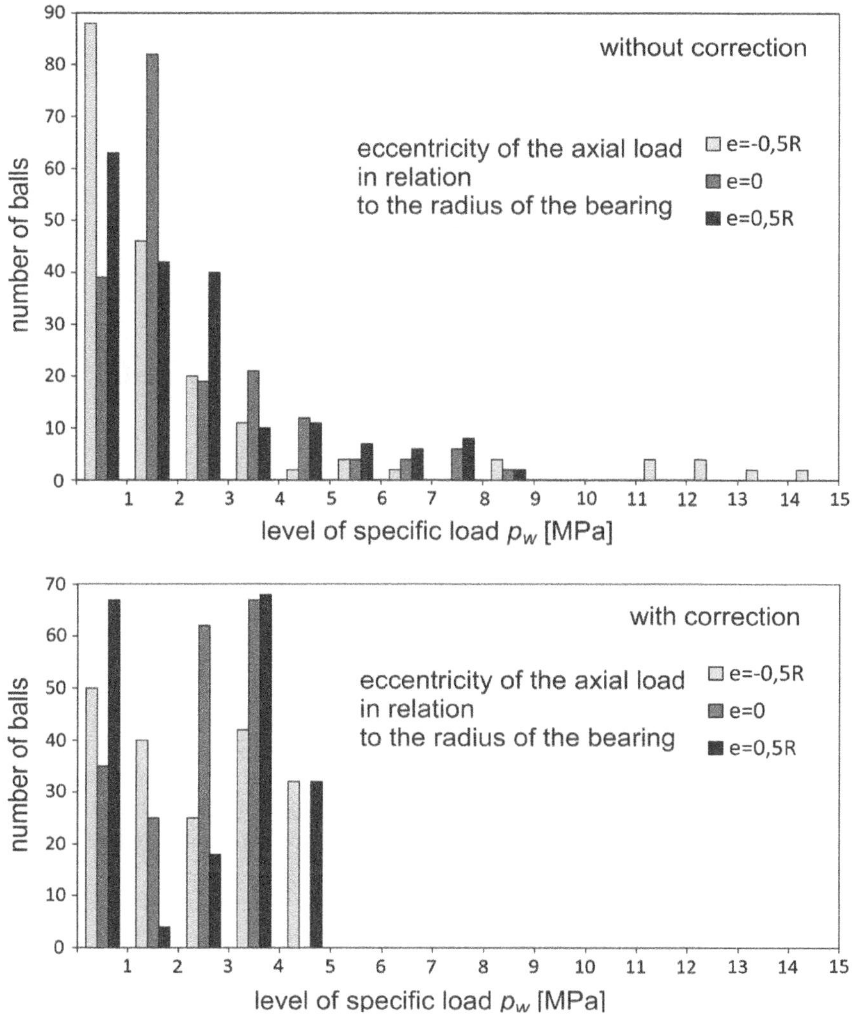

Fig. 16 The comparison of histograms of specific load

The measuring of forces acting upon the balls of the bearing showed that the specific load is equal to about 6.7 MPa. It is a higher value than the value achieved using numerical methods. It results from assembly imperfections and manufacturing errors made during the production of support subassemblies.

10 years after the application of the correction, no plastic wear of raceways was noted. The bearing of a machine that is currently being built will be corrected in an analogous way. A similar correction was applied in the year 2013 to KWK 1500.1 bucket wheel excavator's bearing with a pitch diameter of 8 m.

5 The Measurement of the Load Carried by Rolling Elements of a Bearing

In order to verify computational methods, as well as evaluate the impact of such factors as errors in the manufacturing of support subassemblies, plastic wear of raceways, and correction of raceway's geometry on the distribution of loads in the bearing, two original methods for measuring the load carried by rolling elements of a bearing were developed: a direct method and an indirect method.

5.1 The Indirect Method of Evaluating the Load Carried by Rolling Elements

The method's principle of operation is based on measuring the strain of the support subassemblies' web directly under the raceway of the bearing (Fig. 17).

The contact zone between a rolling element and the raceway has a much smaller dimension than the diameter of a rolling element if measured along the circumferential direction. For that reason, it can be accepted that all forces transmitted by rolling elements are transferred onto the support subassembly (consisting of a ring girder and a raceway) in a quasi-discrete manner. According to Saint-Venant's Principle, the strains in close proximity to the point of application of concentrated forces are not even yet (Fig. 18). The phenomenon described above was utilized to evaluate the load carried by individual rolling elements based on the graph of loads measured during the process of moving them over the point of measurement. Due to the fact the adjacent rolling elements interfere with the measuring signal, this method is only an estimation.

It is necessary to perform calibration. It can be done with the use of FEM model. Once the nodes of the bearing's raceway in the zone of measuring points are loaded

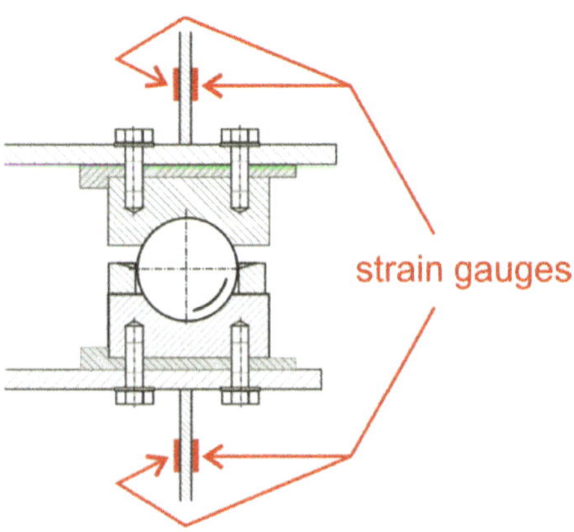

Fig. 17 The principle of measuring the strain of the web directly under the raceway

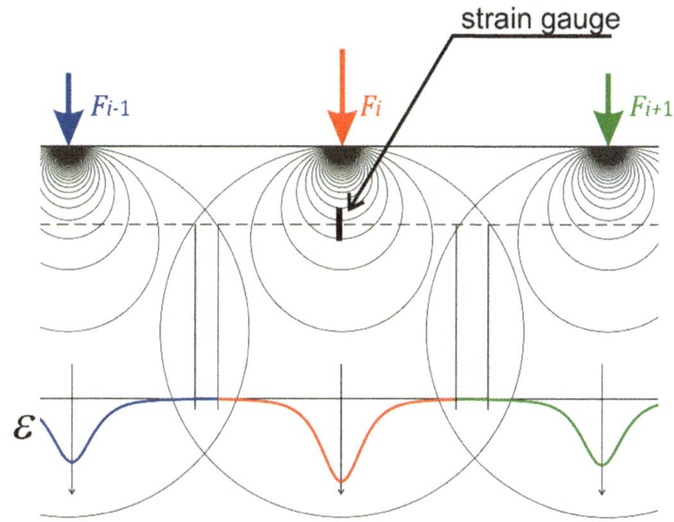

Fig. 18 Strain of the elastic half-space caused by concentrated forces distributed according to the pitch of the bearing

with unit forces, which move around the bearing's circumference relative to the measuring point, the value of the web's strain corresponding to these forces is evaluated. By comparing the measured value of strain pulsation to strain calculated with the use of FEM model, the load carried by an individual rolling element is evaluated (Fig. 19).

It is crucial to properly distribute the gauges. They should be placed as close to the raceway as possible, and as far from the other elements that might interfere with

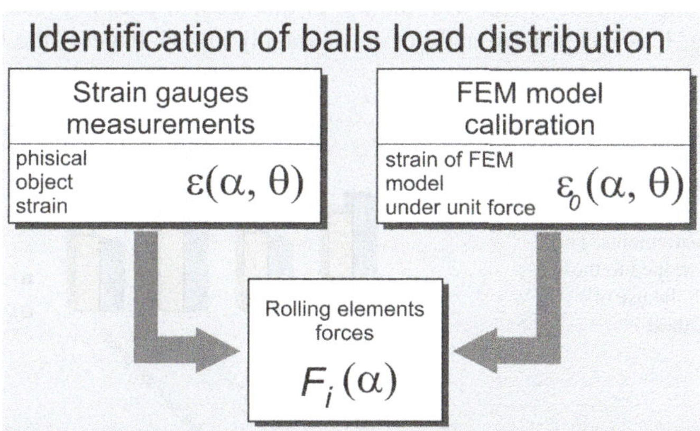

Fig. 19 Identification of the loads carried by rolling elements based on the support subassemblies' strain

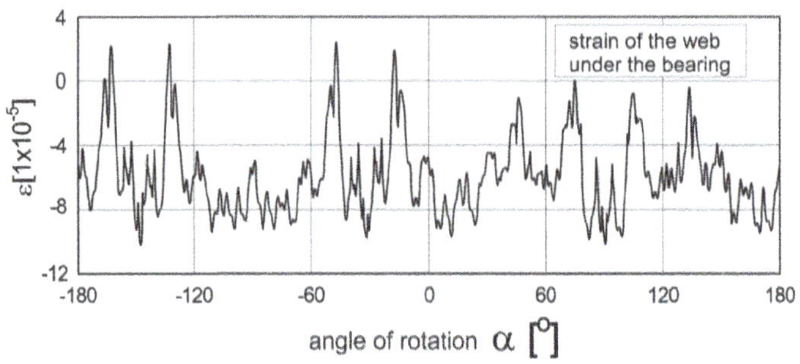

Fig. 20 A graph presenting the strain in the web during the rotation of the superstructure relative to the position of the undercarriage

the field of strains as possible. Placing strain gauges on both sides of the web will eliminate the impact of its bending. This method was described in more detail in works [7, 8].

An example of a graph of the strain of the portal frame's web of the undercarriage is shown in Fig. 20. The difference of strain between the local maximum and the local minima separating it is substantial. The average value of the load carried by rolling elements located in the zones of high structural stiffness is lower than the value evaluated with FEM. However, the average value for adjacent points located in the zones of lower stiffness (between diaphragms) is lower. The results are compared to those evaluated with the use of Ohnrich's model in Fig. 21. The value of the load carried by rolling elements evaluated with the use of Ohnrich's method was set to 100%.

The acquired values of loads confirm the congruity of the results (their graphs and maxima are very similar) evaluated numerically and those obtained by measuring strain pulsation.

Conducting a measurement with the use of this method does not interfere with the bearing. It is inexpensive and possible to apply at any stage of the bearing's

Fig. 21 A comparison of the maximum values of the load carried by balls acquired with the use of FEM and with the measurements. The values are pertained to those obtained with the use of Ohnrich's method

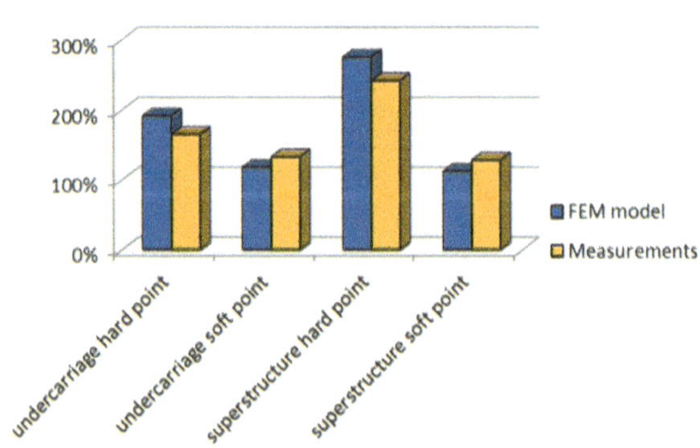

exploitation. The fundamental difficulty is the proper selection of the measuring points. A FEM model of great detail is required in order to do it correctly.

5.2 The Direct Method of Measuring Forces Acting upon Rolling Elements

Another method of measuring the loads carried by rolling elements in bearings was developed by Stańco and Smolnicki [9, 10] during their cooperation with the company KOPEX-Famago in building a new bearing for ŁZKS 1600.33,5 stacker-loader exploited at Bełchatów surface mine (Fig. 22). The bearings of these machines were subject to accelerated wear. The aim of the measurement was to evaluate the level of loads carried by the bearing's rolling elements. The tested bearing had a pitch diameter of 4.485 m and contained 100 balls with a diameter of 110 mm. The raceway of the bearing was made from Ck45 steel in a normalized state. The weight of the superstructure equaled to 560 Mg.

8 sensors with a diameter of the measuring zone equal to 40 mm were built into the bearing in selected locations of the bearing's rings. These sensors had been equipped with strain gauges and were calibrated using a testing machine. The schematic diagram and the appearance of the sensor are presented in Fig. 23. The rolling surface of the bearing was treated together with the bearing's raceway.

An example graph of forces registered by two selected sensors while the superstructure turned left and right is shown in Fig. 24. A good repeatability of measurements was achieved. The statistical error did not exceed ±8% at 95% confidence interval. The results confirmed the overload of the balls in the 'hard' zones and a lack of load in the 'soft' zones.

The measurements confirmed the significant overload of these bearing's rolling elements. The maximum measured value of the specific load carried by a ball equalled 21.7 MPa, which is two times more than the acceptable value for the raceway's

Fig. 22 ŁZKS 1600.33,5 stacker-loader at Bełchatów surface mine and the schematic of the bearing rotating the superstructure

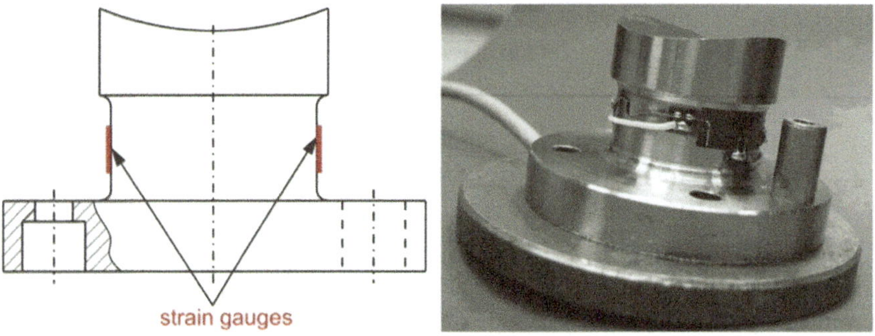

Fig. 23 The sensor for measuring the loads carried by rolling elements: schematic and appearance

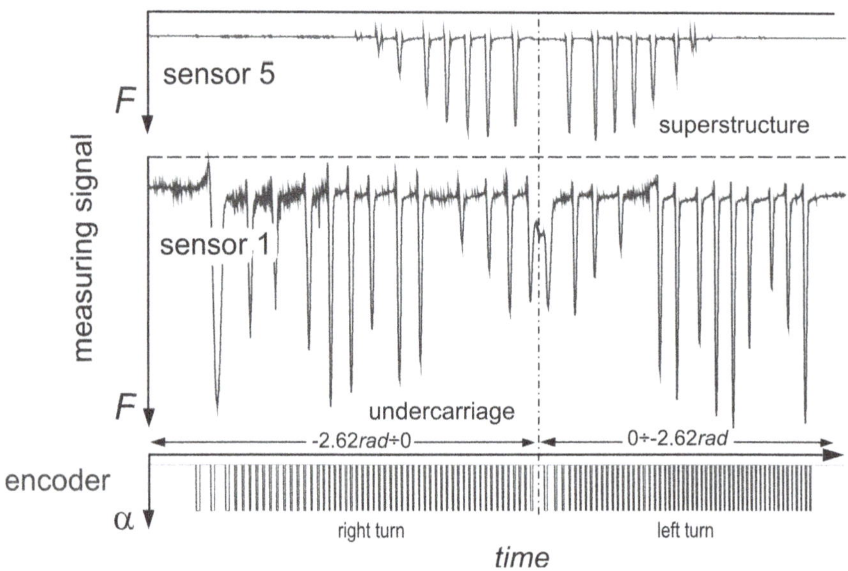

Fig. 24 The graph of measuring signal registered during the rotation of the superstructure

material. By using FEM, a value of the specific load equal to 31 MPa was evaluated in a place into which a sensor could not be built.

Similar measurements were conducted in the year 2007 for the ZGOT 11500.100 stacker's bearing with a pitch diameter of 10 m, into which 13 sensors were built. The advantage of this method is a direct measurement of the load carried by rolling elements and an easy interpretation of results. The disadvantage is the significant cost of producing the sensors which have to be built into the raceway and which greatly increase the production cost of a bearing. It is possible to use them only in new bearings.

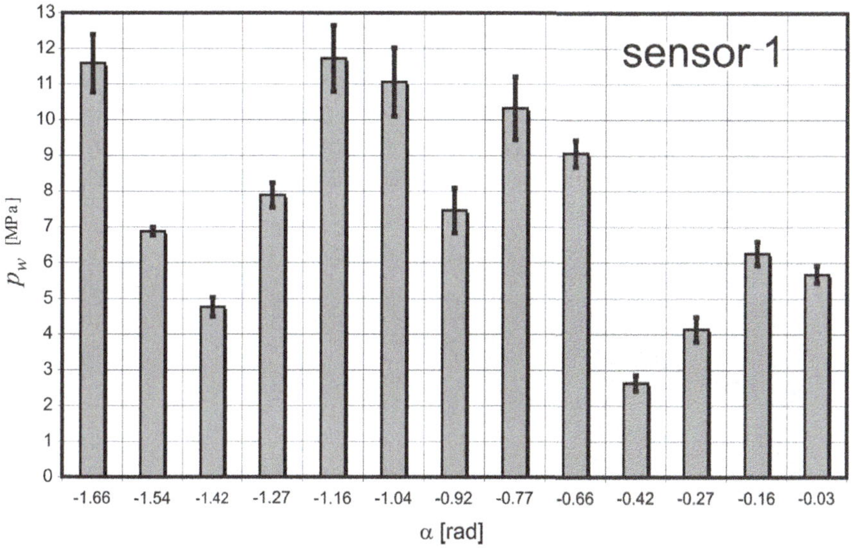

Fig. 25 Specific load carried by a ball—sensor 1–95% confidence interval is marked

If the sensors are built into a raceway made from steel of smaller value of hardness, they are flattened during the exploitation and cease to fulfil their function (Fig. 25).

6 Plastic Wear—Identification and Forecasting

Large-size slewing bearings are prone to frequent damages due to extremely heavy loads carried by rolling elements and the application of raceways made from normalized or heat-treated steel of smaller value of hardness. In case of these bearings, the most fundamental mechanism of wear is the plastic wear. The way plastic wear works is shown in Fig. 26. The material that the raceway is made from flows out

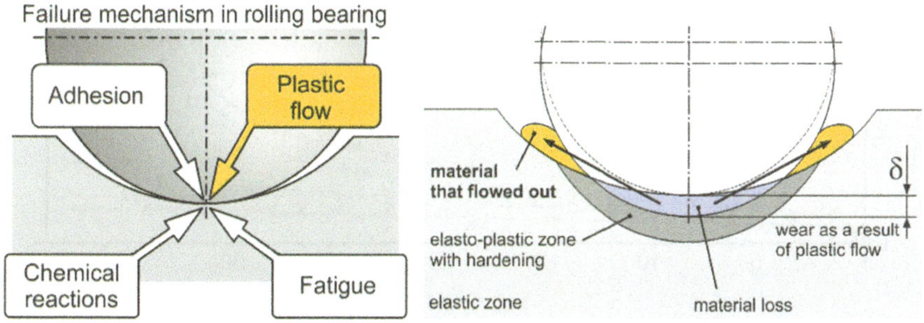

Fig. 26 Basic mechanism of bearing destruction and the mechanism of plastic wear

plastically, which causes the lowering of the raceway, creates ruts and strengthens the surface layer of the raceway through crumpling. Figure 27 presents a raceway of a damaged bearing.

The measurements of these bearing's wear were conducted at Bełchatów surface mine. Figure 28 presents a typical process of plastic wear. It takes place in three phases. After the initial flattening to the value of about 0.5 mm (phase I), a slow

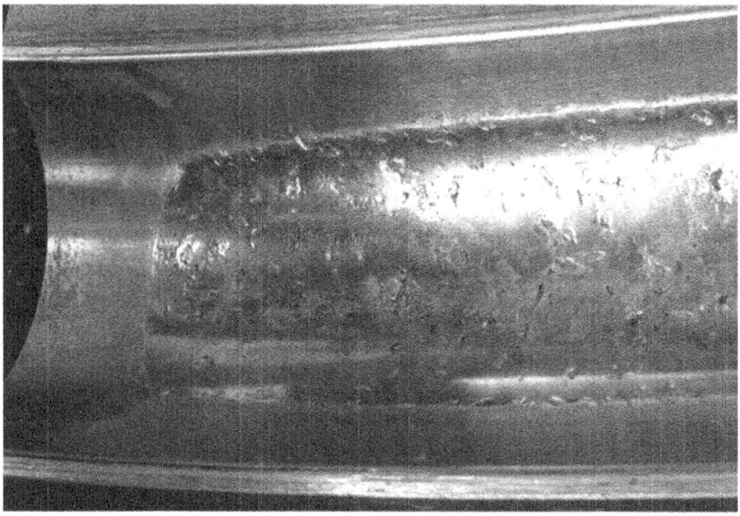

Fig. 27 A degraded, normalized raceway of a stacker-loader's bearing made from steel Ck45 after 4 years of exploitation, cause: the support subassembly was not stiff enough

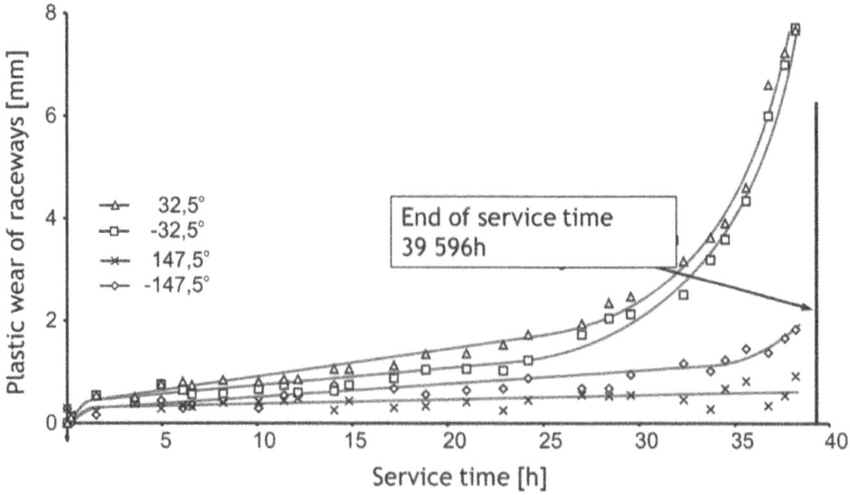

Fig. 28 Plastic wear of stacker-loader's raceway over time

increase of wear takes place up to the value of 2 mm (phase II). Having reached the said value of wear, the accelerated degradation of the bearing takes place (phase III), until the bearing is withdrawn from exploitation. An autogenous correction of the bearing's support deviation takes place in phase I. In phase II (the stable part of the graph), plastic wear is still the most fundamental mechanism of wear. The speed of wear growth can reach up to 55 μm for every 1000 h of work. In phase III, the speed increases to up to 0.9 mm for every 1000 h. Numerous loosenings of the raceway's surface material of fatigue character take place during this phase. The material that flowed out is very often flattened on the sidewalls of the raceway. New structures of material of layered character then appear, creating a secondary raceway [11]. An example of such structure is shown in Fig. 28. The identification of the new, deformed raceway is a separate problem, which is described in an article [12] and in a chapter of a monograph [7] (Fig. 29).

Phase I is often intensive in large bearings. Apart from correcting the geometrical deviations of the support subassemblies, it also corrects their uneven distribution of stiffness. Such a situation took place in a stacker's bearing with a diameter of 10 m, containing 188 balls with a diameter of 150 mm. The load distribution of the stacker's bearing as shown in Fig. 7. Plastic wear of each raceway was measured in 4 for points after 5 years of service [14]. The value of wear reached almost 5 mm (Fig. 30). An analysis of the form of the wear and its impact on load distribution was conducted. Figure 31 presents the form of load distribution of a new bearing, the form of load distribution of a new bearing with the flatness deviation of supporting structure factored in, and the form of load distribution of a bearing exploited for 5 years [15].

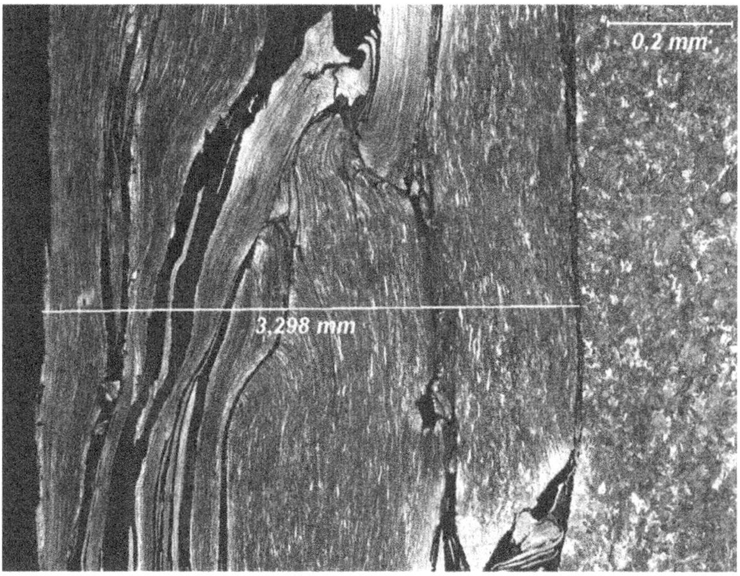

Fig. 29 Structure of a secondary raceway [13]

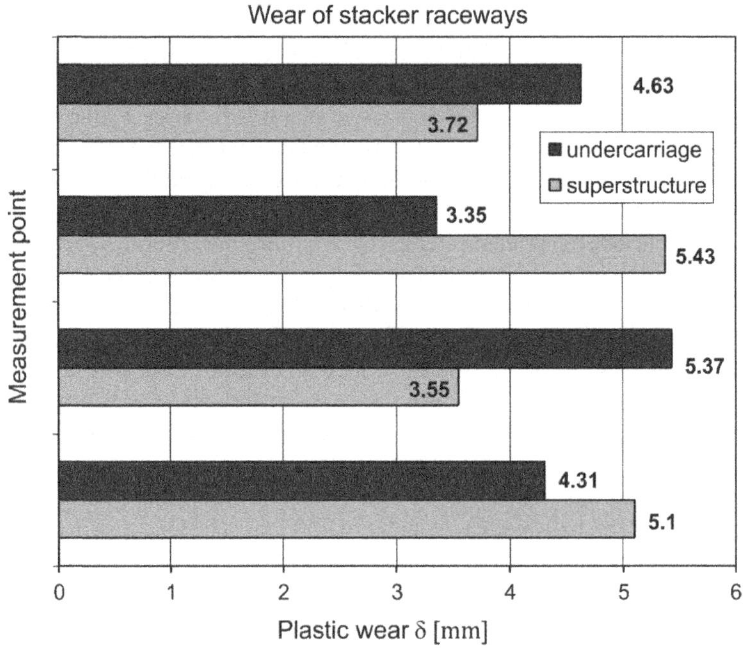

Fig. 30 Plastic wear of stacker's raceways after 5 years of service

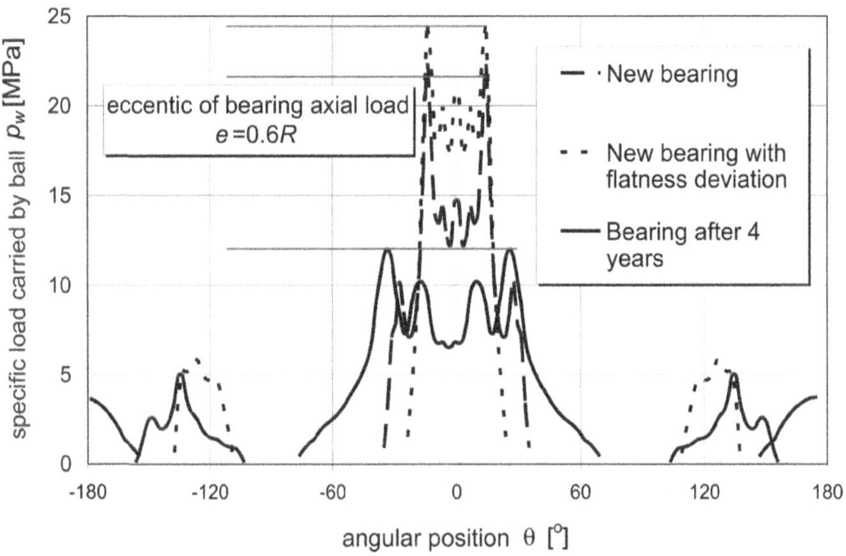

Fig. 31 A comparison of specific load carried by a new bearing and a bearing exploited for 5 years

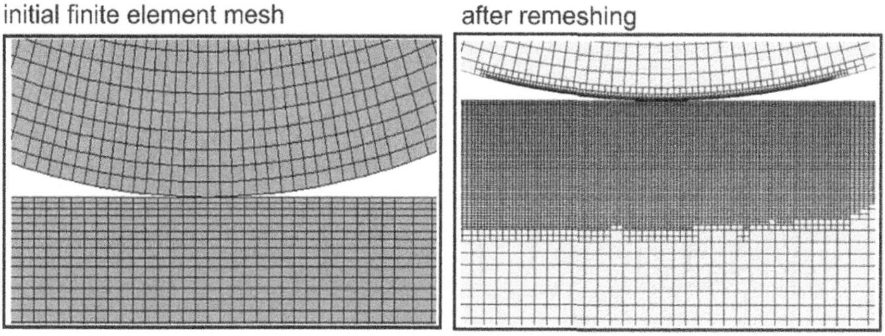

Fig. 32 The application of adaptive mesh to simulate plastic wear

A significant decrease in the maximum value of the specific load carried by rolling elements took place, decreasing from 24.8 to 12.3 MPa. A 5-year forecast of safe exploitation of the bearing was formulated based on the analysis. Further service confirmed the predictions. Plastic wear stopped and the bearing was in use for the next 7 years until the scrappage of the machine.

Forecasting the speed of plastic wear growth requires time-consuming experimental research. However, if the model of the material is well known, including the way in which the material was strengthened, it is possible to evaluate the number of cycles of rolling a ball over-plastic wear characteristics with the use of FEM numerical simulations. It requires creating models with multiple degrees of freedom and powerful computers.

First attempts at analyzing plastic wear while factoring rolling over (rolling over means a simulation of a full revolution of the machine's superstructure conducted on a FEM model) in were made by Stańco [9, 16], who utilized plane strain 2D elements (Fig. 32). Due to high values of plastic deformations, it was necessary to use adaptive mesh and remeshing technique. The results confirmed the stabilization of plastic wear's speed during the first cycles of loading (Fig. 33). The solid line on the graph indicates the way the deflection of a raceway changed with subsequent roll overs, while the dashed red line illustrates the plastic deflection of irreversible character. Due to simplifications, these models were only qualitative in character.

Nowadays, mostly 3D models are created, which require computers with high processing power. Figure 34 presents the initial-boundary conditions applied to the model. A very dense discretization with the use of finite elements was applied to the gray area. Every cycle consists of: loading the model with force, rolling the raceway over back and forth, and unloading the model.

The model of the material was evaluated based on samples of the 42CrMo4 V steel in its normalized state collected from a ring of a worn-out bearing outside of its zone of plastic wear. The equivalent plastic strain was obtained as a result.

Figure 35 presents the value of equivalent plastic strain after 1, 2, 7, and 14 cycles of rolling a ball over on the raceway. A high increase of wear takes place initially; however, between the cycles 7 and 14, it becomes significantly lower. It is even more

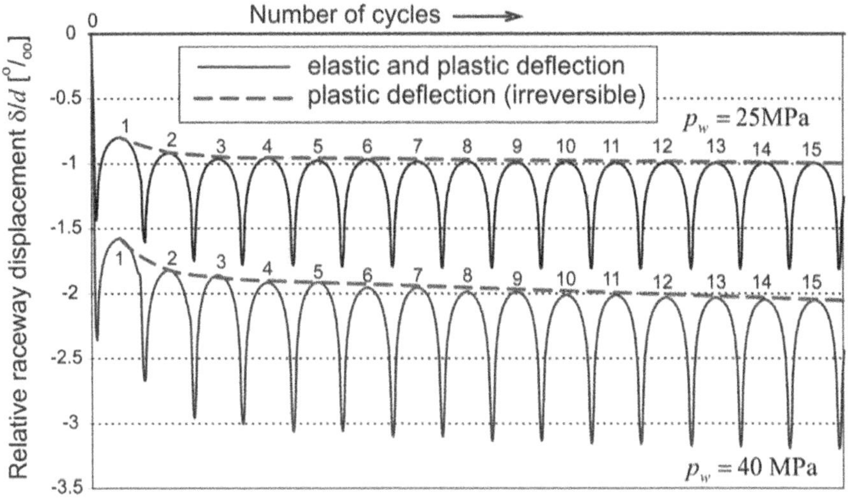

Fig. 33 Elastic and plastic raceway deflection at different values of specific load

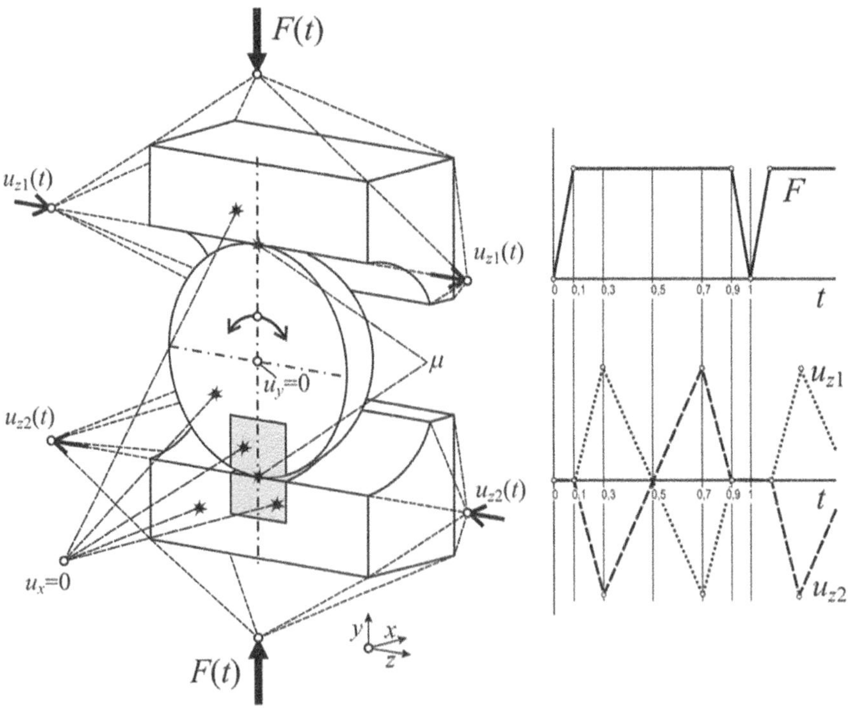

Fig. 34 Schematic of a 3D model used for simulating plastic wear with boundary conditions

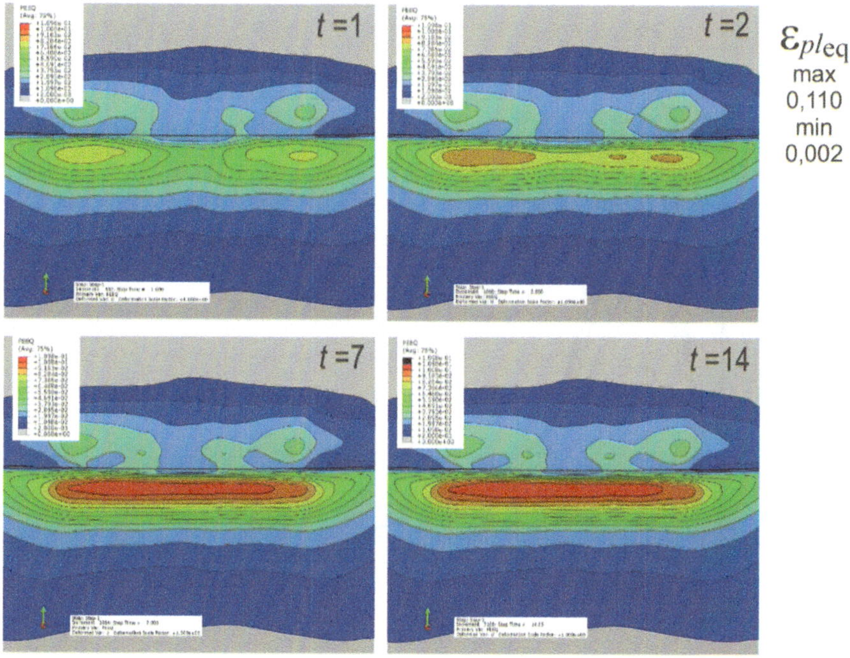

Fig. 35 Equivalent plastic strain after 1, 2, 7, and 14 cycles of rolling over—longitudinal section of the raceway

visible on the graph shown in Fig. 36. A concentration of plastic deformations in the Bielaiev's zone at the depth of between 0.3 and 2.5% of the diameter of a rolling element is visible. In the zone above the 3% of the diameter of a rolling element, the plastic deformation took place only during the first cycle.

The calculations were repeated for different values of the specific load carried by a rolling element. The results of plastic deflection of a raceway after n cycles of loadings are shown in Fig. 36. The higher the value of the specific load, the faster the increase of wear. The model of plastic wear was evaluated based on these curves. The equation of the model is:

$$\delta = \alpha \times n + b \tag{18}$$

and it contains a constant component b, which causes the wear in the initial phase, and a linear component $\alpha \times n$, which causes an increase in the stable phase (Fig. 37). The value of α is evaluated based on the tangent of the slope in the stable phase. The values of constants α and b depend on the material the raceway is made from and the value of the specific load. The constant α depends linearly on the value of specific load in the analyzed range (Fig. 38).

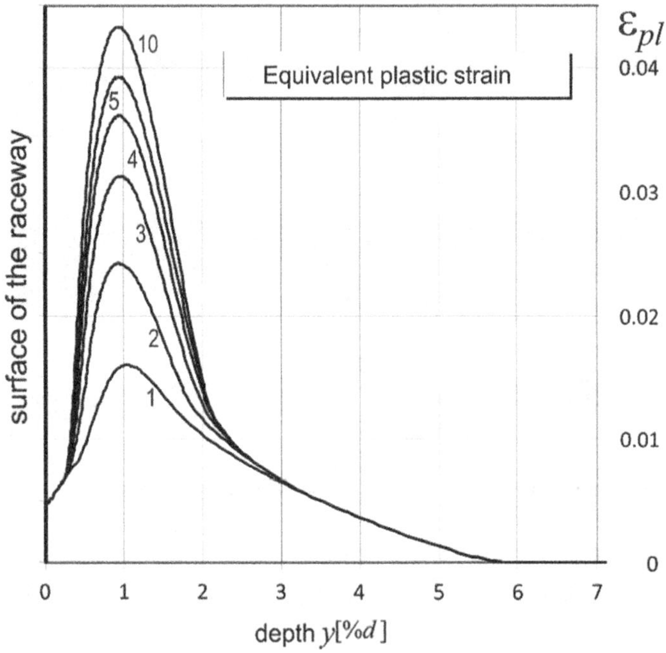

Fig. 36 Distribution of equivalent plastic strain inside the raceway after 1, 2, 3, 4, 5, and 10 cycles of rolling over

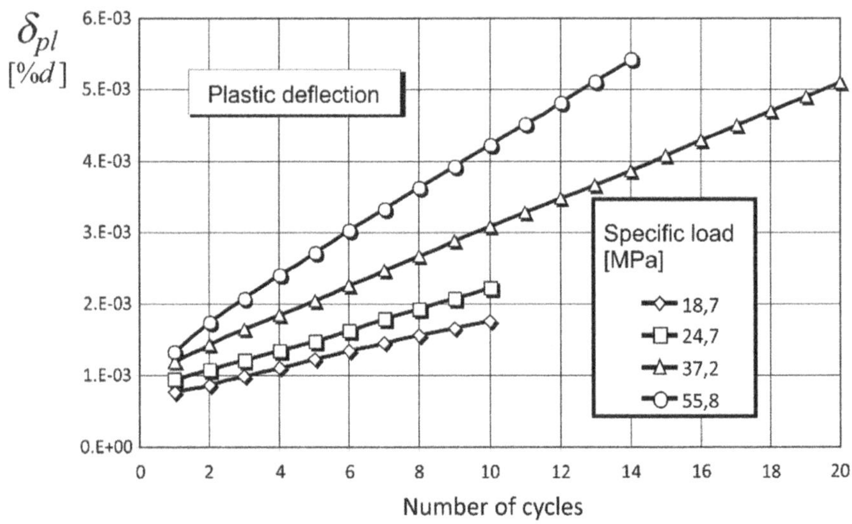

Fig. 37 Plastic deflection at different values of specific load

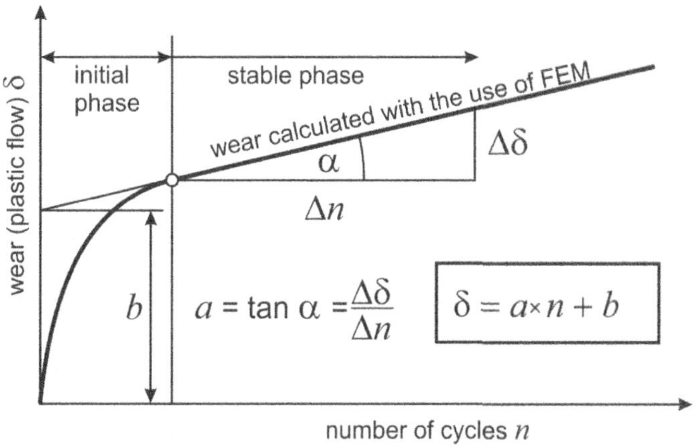

Fig. 38 Plastic wear obtained with the use of a numerical model

7 Summary

The selected information about the research conducted at the Faculty of Mechanical Engineering at Wroclaw University of Science and Technology deal with a whole spectrum of issues concerning large-size rolling bearings, which mostly occur as slewing bearings of all sorts.

The developed computational models of the raceway-rolling element-raceway system based on FEM are nowadays utilized by scientists in many countries [17–31]. At the same time, they are the basis for cooperation with heavy equipment and extractive industries. Machines such as stackers and bucket wheel excavators are exploited even for 50 years and not only do they create a good groundwork for cooperation, but they also make for a great research material. For the needs of the industry, the method of correcting raceways, which decreases the load carried by rolling elements almost by half, was developed. It has already been applied to three machines and will be applied to yet another.

The developed indirect method of evaluating the load carried by rolling elements based on strain pulsation in the web under the bearing's ring is easy to apply. The measurements of the load carried by rolling elements conducted with the use of sensors built into the raceway are the only measurements of this kind in terms of large-size ball bearings. They confirmed the correctness of the created computational FEM models.

Numerous research papers concerning plastic wear result from using normalized steel or heat-treated steel of smaller value of hardness for building the rings of such bearings. Plastic wear, unlike in the case of typical, catalogue slewing bearings, is the most dominant mechanism of destruction in these bearings, which is dominated by fatigue wear only in the final phase. It is currently planned to carry out experimental research aimed at fine-tuning the models of plastic wear for the typical materials used for these bearings and evaluating the acceptable plastic deformation of the raceway's material at which the intensification of fatigue mechanisms takes place.

References

1. Nogieć, T., Malcher, K.: Rozwój teorii i metod obliczeń oraz kształtowania łożysk wielkogabarytowych. Etap1. (Development of the theory and methods of calculation and shaping of large-size bearings. Stage 1). IKEM report of the Wroclaw University of Technology No. 86/S-043, Wroclaw 1986 (in Polish)
2. Nogieć, T., Smolnicki, T.: Opracowanie metod i urządzeń do doświadczalnej oceny wpływu parametrów konstrukcji i obciążeń na nośność i trwałość łożysk wieńcowych. (Development of methods and equipment for experimental evaluation of the influence of construction parameters and loads on the load capacity and durability of ring bearings). IKEM report of the Wroclaw University of Technology Ser. SPR. No. 111, Wroclaw (1987) (in Polish)
3. Smolnicki, T., Rusiński, E.: Superelement-based modeling of load distribution in large-size slewing bearings. J. Mech. Des. **129**(4), 459–463 (2007)
4. Smolnicki, T.: Nieliniowe modele układu bieżnia-kula-bieżnia do wyznaczenia rozkładu nacisków w wielkowymiarowym łożysku tocznym. (Non-linear track-to-ball track arrangement models to determine the pressure distribution in a large dimensional rolling bearing). Przegląd Mechaniczny R. 58, nr 5/6, 16–20 (1999) (in Polish)
5. Smolnicki, T., Przybyłek, G.: Wpływ korekcji dźwigara pierścieniowego na dystrybucję obciążenia w wielkogabarytowym łożysku tocznym. Metody doświadczalne w budowie i eksploatacji maszyn. (Effect of ring girder correction on load distribution in large rolling bearing. Experimental methods in machine construction and operation). Vth Scientific Conference, Wrocław-Szklarska Poręba, [14–16 maja] T. 2. Wroclaw, 209–216 (2001) (in Polish)
6. Przybyłek, G.: Metoda uzyskiwania równomiernego obciążenia w parach obrotowych o niejednorodnej podatności. (Method for obtaining a uniform load in rotary pairs of heterogeneous susceptibility). Ph.D. thesis. IKEM report of the Wrocław University of Technology. Ser. PRE No. 14 (2003) (in Polish)
7. Smolnicki, T.: Fizykalne aspekty koherencji wielkogabarytowych łożysk tocznych i odkształcalnych konstrukcji wsporczych. (Physical aspects of coherence of large-size rolling bearings and deformable support structures). Monograph. Publishing House Oficyna Wydawnicza Politechniki Wroclawskiej, Wroclaw (2002) (in Polish)
8. Smolnicki, T.: Numerical-experimental method of identifying load distribution in large-size bearings. In: 22nd Danubia-Adria Symposium on Experimental Methods in Solid Mechanics. Extended abstracts. Italian Association for Stress Analysis, Monticelli Terme-Parma, Italy, September 28–October 1, 274–275 (2005)
9. Stańco, M.: Modele analityczno-numeryczne zużycia odkształceniowego wielkogabarytowych łożysk tocznych. (Analytical and numerical models of deformational wear of large-size rolling bearings). Ph.D. thesis. IKEM report of the Wrocław University of Technology Ser. PRE No. 2, Wroclaw (2008) (in Polish)
10. Smolnicki, T., Stańco, S., Pietrusiak, D.: Distribution of loads in the large size bearing—problems of identification. Tehnički Vjesnik—Technical Gazette **20**(5), 831–836 (2013)
11. Smolnicki, T., Pękalski, G., Jakubik, J., Harnatkiewicz, P.: Investigation into wear mechanisms of the bearing raceway used in bucket wheel excavators. Arch. Civ. Mech. Eng. **17**(1), 1–8 (2017)
12. Smolnicki, T., Koziołek, S.: Methods of identifying plastic wear of large-size bearing tracks. W: Mechatronic systems and materials, MSM 2011: 7th international conference, 7–9 July 2011, Kaunas (Lithuania) (2011)
13. Smolnicki, T., Harnatkiewicz, P., Stańco, M.: Degradation of a geared bearing of a stacker. Arch. Civ. Mech. Eng. **10**(2), 131–139 (2010)

14. Smolnicki, T.: Large—size bearings in opencast mining machines. In: Bhattacharya, J. (ed.) Design and Selection of Bulk Material Handling Equipment and Systems: Mining, Mineral Processing, Port, Plant and Excavation Engineering, vol. 1, pp. 105–130. Wide Publishing, Kolkata (2012)
15. Smolnicki T., Stańco M.: Zmiana obciążeń elementów tocznych w łożu kulowym zwałowarki wskutek zużycia odkształceniowego (Change of loads of rolling elements in the bearing raceway of a stacker as a result of deformational wear). Górnictwo i Geoinżynieria R. 35, z. 3/1, 239–245 (2011) (in Polish)
16. Smolnicki, T., Stańco, M.: Prognozowanie zużycia odkształceniowego wielkogabarytowych łożysk tocznych o bieżniach miękkich. (Forecasting of deformational wear and tear of large-size roller bearings with soft raceways). Acta Mechanica et Automatica **3**(1), 98–100 (2009). (in Polish)
17. Ohnrich, S.: Berechnung der zweireihigen Kugeldrehverbindungen. Institut für Fördertechnik, Leipzig (1959)
18. Mathias, K.: Berechnung der Wälzkörperkräfte in Grosswälzlagern. Fördern und Heben 29/1979
19. Smolnicki, T., Derlukiewicz, D., Stańco, M.: Evaluation of load distribution in the superstructure rotation joint of single-bucket caterpillar excavators. Autom. Constr. **17**(3), 218–223 (2008)
20. Yu, Y.H., Lee, B.R., Cho, Y.J.: New load distribution method for one-row slewing ball bearing considering retainer force. Int. J. Precision Eng. Manufact. **18**(1), 49–56 (2017). https://doi.org/10.1007/s12541-017-0006-7. Published: Jan
21. Abasolo, M., Coria, I., Plaza, J., Aguirrebeitia, J.: New selection curves for four contact point slewing bearings. Proc. Inst. Mech. Eng. Part C-J. Mech. Eng. Sci. **230**(10), 1715–1725 (2016). https://doi.org/10.1177/0954406215583522. Published: Jun
22. Aithal, S., Prasad, N.S., Shunmugam, M.S., Chellapandi, P.: Effect of manufacturing errors on load distribution in large diameter slewing bearings of fast breeder reactor rotatable plugs. In: Proceedings of the Institution of Mechanical Engineers Part C-Journal of Mechanical Engineering Science, vol. 230, Issue 9 (Special Issue: SI), pp. 1449–1460 https://doi.org/10.1177/0954406215579947 Published: May 2016
23. Goncz, P., Ulbin, M., Glodez, S.: Computational assessment of the allowable static contact loading of a roller-slewing bearing's case-hardened raceway. Int. J. Mech. Sci. **94–95**, 174–184 (2015). https://doi.org/10.1016/j.ijmecsci.2015.03.006. Published: May
24. Chen, G.C., Wen, J.M.: Effects of size and raceway hardness on the fatigue life of large rolling bearing. J. Mech. Sci. Technol. **29**(9), 3873–3883 (2015). https://doi.org/10.1007/s12206-015-0833-3. Published: Sep
25. Kania, L., Krynke, M., Mazanek, E.: A catalogue capacity of slewing bearings. Mech. Mach. Theory **58**, 29–45 (2012). https://doi.org/10.1016/j.mechmachtheory.2012.07.012. Published: Dec
26. Hai, G.X., Diao, H.X., Jing, H.R., Hua, W., Jie, C.: A rolling contact fatigue reliability evaluation method and its application to a slewing bearing. J. Tribol.-Trans. ASME **134**(1) Article Number: 011101. https://doi.org/10.1115/1.4005770. Published: Jan 2012
27. Qin, S.X., Li, Y.L.: Stress analysis of the working device of multi-functional excavator. Adv. Manuf. Technol. PTS 1–3 Book Series: Adv. Mater. Res. **314–316**, 2499–2503. https://doi.org/10.4028/www.scientific.net/AMR.314-316.2499. Part: Part 1-3 Published: 2011
28. Liu, H.B., Li, J.S., Xue, Y.J., Ma, W.: Study on Load distribution calculating method of slewing bearing with the link elements. Adv. Sci. Lett. **4**(8–10), 2759–2763 (2011). https://doi.org/10.1166/asl.2011.1638. Published: Aug–Oct
29. Spiewak, S.: Methodology for calculating the complete static carrying capacity of twin slewing bearing. Mech. Mach. Theory **101**, 181–194 (2016). https://doi.org/10.1016/j.mechmachtheory.2016.03.017. Published: JUL
30. Starvin, M.S., Manisekar, K.: The effect of manufacturing tolerances on the load carrying capacity of large diameter bearings. Sadhana-Acad. Proc. Eng. Sci. **40**(6), 1899–1911 (Published: Sep 2015)

31. Lacroix, S., Nelias, D., Leblanc, A.: Four-point contact ball bearing model with deformable rings. J. Tribol.-Trans. ASME **135**(3) Article Number: 031402, https://doi.org/10.1115/1.4024103. Published: Jul 2013

Problems of Noise Hazards in Long-Term Operated Basic Machinery for Open-Pit Mining

Piotr M. Sokolski and Marek Sokolski

Abstract Long-term operated open-pit mining machinery (large-size excavators and stackers) are subjected to some various degradation processes. One of the symptoms of these phenomena is increased noise levels which have a negative influence on a microclimate of work environment of machines' crews. That is why since the 1990s the complex program of modernization of all the old basic machinery has been consistently put into practice in Polish brown coal industry and the comprehensive monitoring of noise hazards in this machinery is consistently being carried out. In this chapter, the main reasons and sources of increased noise are identified and the relevant research results obtained by authors are presented.

Keywords Long-term operated machinery · Noise hazards

1 Introduction

Brown coal is one of two main energy resources in Poland and about 40% of domestic electric energy production is based on this fuel. Brown coal is getting out by open-pit mining technology with the use of large-size machinery: bucket-wheel excavators, bucket-chain excavators and stackers which are usually operated 24 h a day, 7 days a week. That is why these machines are particularly subjected to various degradation processes which contribute to the reduction of safety level and in extreme cases it can cause dangerous events: serious damages or even catastrophes [1, 2].

One of the symptoms of degradation processes occurring in this long-term operated mining machinery is increased noise which has a quite negative influence on the overall conditions in the work environment [3, 4].

The large-size mining machinery in lignite mines is notorious for destroying human hearing. In the past mineworkers had been the only people to be affected. Now, with pits being closer to inhabited areas, outsiders are becoming affected as well.

P. M. Sokolski (✉) · M. Sokolski
Faculty of Mechanical Engineering, Wroclaw University of Science and Technology, ul. Wybrzeze Wyspianskiego 27, 50-370 Wroclaw, Poland
e-mail: piotr.sokolski@pwr.edu.pl

M. Sokolski (ed.), *Mining Machines and Earth-Moving Equipment*,
https://doi.org/10.1007/978-3-030-25478-0_4

These are additional reasons why these old machines should be successively subjected to overhaul. Since the 1990s the complex program of modernization of all the old basic machines has been consistently put into practice in Polish brown coal companies.

2 Origins, Causes, and Features of Noise Hazards in Large-size Open-pit Machines

Depending on the size of the affected zone, noise processes occurring in large-size mining machinery have a significant impact on:

- local acoustic climate in a relative small space, e.g., in operator's cabins, in engine rooms, in crew's social rooms, on gangways and service platforms, etc.,
- regional acoustic climate in some larger space, e.g., in the neighborhood of a machine or in the vicinity of open-pit.

The main sources of noise hazards in large-size machinery for open-pit mining are located first of all in the areas of drive units, band conveyors as well as loading and discharging zones. The principal causes of increased noise in these areas are highly variable internal and external loads generated in working process.

Internal loads are mainly due to inertia forces arising in rotational motion of unbalanced masses within drive systems. In addition, degradation processes in kinematic pairs result in excessive clearances that usually cannot be corrected by standard maintenance procedures.

External loads are random in character and come in particular from contact between machine tools (buckets) and a rock face or between chutes and falling material (brown coal or overburden). Worn-out buckets are usually not capable of assuring smooth cutting. Even worse, deeper layers of lignite that are excavated at present have more embedded boulders and hard rock inclusions than near-surface seams. In general, the thirty-odd years old large-size excavators were not intended for use under such adverse conditions.

The characteristic features of the acoustic climate in the large-size machinery for open-pit mining are as follows:

- high noise level due to high installed power of drive units,
- long exposure time because these machines are continuously operated,
- limited usability of sound-absorbing shields or screens, because they could hinder to perform required technical inspections.

The noise sources in the large-size mining machinery are located (a) in the open space (e.g., drive units of bucket-wheels or crawler drive mechanisms), in partially confined space (e.g., carrying and return idlers of belt conveyors located along the gangways and service platforms) or (b) in full closed areas (e.g., in operator's cabins, social rooms, engine rooms).

However, noise emitted by a point source located in the open space quickly decreases as the distance from this source increases.

The sound pressure level emitted by a point source of noise located in the open space quickly decreases as the distance from the source increases. This phenomenon can be described by the following simplified formula:

$$L(r_2) \cong L(r_1) - 10\log\frac{r_2^2}{r_1^2} \tag{1}$$

where

r_1 and r_2 denote the radial distance from the noise source,
$L(r)$ is the sound pressure level at radial distance r

The formula (1) shows that each doubling the distance from the source causes the noise level to drop by 6 dB.

Decreasing the acoustic intensity of noise sources in the large-sized mining machinery is a particularly important goal because of the excessive noise:

- can make verbal communication difficult or impossible,
- can mask acoustic warning signals,
- has a negative impact on the hearing organs of operators, causing, e.g., temporary or permanent hearing impairment,
- can cause extra-auditory effects, in the form of disruption of the basic body's physiological functions (occupational diseases).

However, if the noise sources are located in a closed space, then the acoustic waves cannot propagate freely and are reflected on walls and other objects inside. For the assessment of acoustic microclimate in the rooms located in basic machines for open-pit mining (operator's cabins, engine rooms, social rooms, etc.), the following facts are important:

- at a relatively small distance from the sound source, below the limit value r_0, the acoustic activity of the source (the energy of incident waves) has a dominant influence on the total noise level,
- at a greater distance, above the limit value r_0, the acoustic absorption of the room (energy of waves reflected from walls or other internal objects) has a decisive influence on the noise level.

In addition, it should be noted that ventilation and heating devices can significantly affect the acoustic climate in the rooms for the crews of basic machines for open-pit mining. Fans and blowers are sources of infrasound noise in which the components with frequencies $f \leq 20$ Hz are usually dominant, and the threshold of audibility of this noise is above 90 dB. Typical effects in people exposed to prolonged infrasound noise are the following symptoms:

- feeling of pressure in the ears that causes states of excessive fatigue, discomfort, difficulty in concentration, drowsiness, etc.,

- feeling of "internal vibrations"—as a result of dangerous resonance of the body's internal organs, which can finally lead to physiological changes (occupational diseases).

3 "Hot Spots" of Noise in Large-size Open-Pit Machines

In general, they are the following most intensive sources of noise ("hot spots") in large-size bucket-wheel excavators, bucket-chain excavators and stackers:

- the drive unit of a bucket-wheel,
- the drive unit of a bucket-chain and elements of this chain,
- the local discharging zone where material (brown coal or overburden) is falling down from a bucket-wheel or bucket-chain on a belt conveyor mounted on an excavating boom,
- the drive unit of belt conveyor and conveyor idlers installed on excavating boom,
- the central discharging zone where the material is falling down from the first belt conveyor on the second belt conveyor,
- the drive unit of slewing mechanism,
- the drive unit of crawlers and elements of crawler chains.

In these above-mentioned zones, the most important acoustic hazards usually occur and therefore the presence of people (staff) in these places should be limited to the necessary minimum. The high noise level occurred in these "hot spots" results, inter alia, from the large installed drive power.

European Union standards recommend using the sound power emitted by a source as a measure of noise affecting the surroundings. The total level of noise $L_{\mathrm{A(perm)}}$ depends on machine type and its built-in power. According to these guidelines, if the installed power is designated by P (kW), then the values of noise level considered as statistical expected values can be assumed as follows:

$$L_{\mathrm{A}} = 80 + 11 \cdot \log P \tag{2}$$

However, it should be noticed that the places of prolonged human presence, i.e., especially operator cabins, machinery room, and social rooms must be particularly protected against noise and therefore the most restrictive criteria for these zones are valid.

4 Basics of Acoustic Climate Evaluation in Large-size Machines for Open-Pit Mining

A preliminary assessment of the acoustic climate in basic opencast mining machines is generally carried out on the basis of the following quantities:

- the A-weighted equivalent sound pressure level (frequency weighting "A") L_{Aeq} which is determined for a nominal eight hour exposure period,
- maximum sound pressure level (frequency weighting "A") L_{Amax}
- maximum peak sound pressure level (frequency weighting "C") $L_{\mathrm{C,Peak}}$.

The permissible occupational exposure limits of noise level are defined in proper international standards (e.g., ISO R 1996) and in the Ordinance of the Polish Minister of Labor and Social Policy ("Ordinance on maximum permissible concentration and intensity of harmful factors in the work environment in accordance with national limit values"). In accordance with these guidelines, the noise level should not exceed the following values:

- in the workplaces $L_{\mathrm{Aeq(perm)}} = 85$ dB(A),
- in the cabins without telephone communications $L_{\mathrm{Aeq(perm)}} = 75$ dB(A),
- in the cabins with telephone communications $L_{\mathrm{Aeq(perm)}} = 65$ dB(A),
- in the social rooms $L_{\mathrm{Aeq(perm)}} = 55$ dB(A).

However, in any case, the following values of noise level are not allowed to exceed:

- maximum sound pressure level (frequency weighting "A") $L_{\mathrm{Amax}} = 115$ dB(A),
- maximum peak sound pressure level (frequency weighting "C") $L_{\mathrm{CPeak}} =$ 135 dB(C).

The lignite mines are increasingly approaching built-up areas and therefore their residents are affected by industrial noise which can be especially nagging during the night. There are some relevant regulations that determine permissible noise levels, e.g., the regulations issued by the Polish Ministry of Environment, Natural Resources and Forestry (see Government Journal of Bills 98.66.435, June 1, 1998) stipulates that the noise affecting areas with detached housing should not exceed 40 dB(A) during nights (Dudek and Sokolski).

The more detailed assessment of acoustic microclimate, according to the recommendation of ISO R 1996, is performed on the basis of the sound spectrum and so-called Noise Rating curves *NRx*. According to this, the octave noise spectrum is compared with the values of the relevant curve *NRx*, recommended for a given zone where the noise hazards are assessed (e.g.,, in cabs, engine rooms, etc.). Index x of noise curve *NRx* is assumed from the following formula:

$$x \cong L_{\mathrm{Aeq,8h}} - 5;\ [\mathrm{dB(A)}] \tag{2}$$

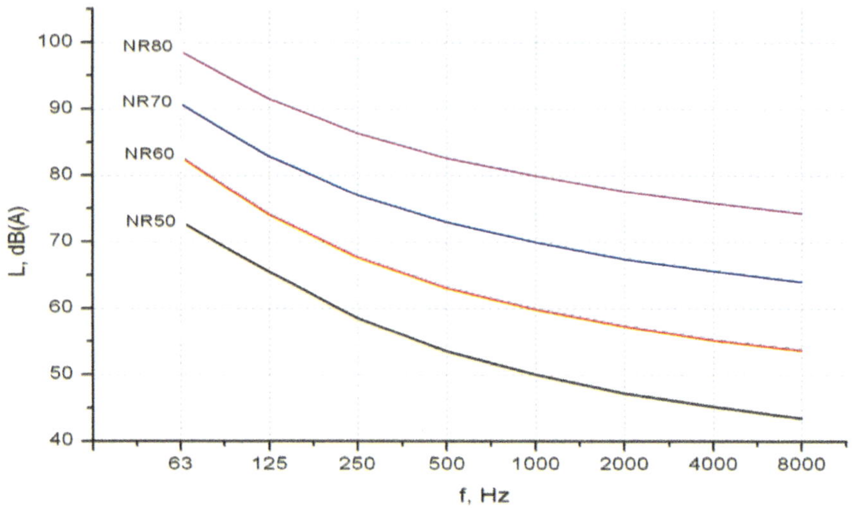

Fig. 1 Noise rating curves *NRx* for assessment of acoustic microclimate in large-size mining machines

where:

$L_{\mathrm{Aeq,8h}}$ denotes the A-weighted equivalent sound pressure level which is permissible for the sake of ability to perform basic operations by worker

The Noise Rating Curves *NRx* proper for the assessment of acoustic microclimate in the previously mentioned places are presented in Fig. 1.

5 Case Studies of Noise Hazards

As examples, there are several cases of assessment of acoustic hazard that are characteristic for large-size mining machinery for brown coal open-pit mining.

The sound level generated in these "hot spots" of noise is a random variable. Therefore, noise hazards should be treated in a probabilistic approach. Such an approach was performed in the authors' work [5].

With this in mind, the assumption was made that the noise level in the analyzed zones has a normal statistical distribution. Next, the upper values of 95% confidence intervals on mean values were assumed as representative values of the total noise level.

5.1 Case Study#1—The Noise Hazards in the Zones of Main Drive Units in Bucket-Chain Excavators

Old long-term operated bucket-chain excavators belong to the most acoustically active basic machines in open-pit mines. This is particularly true for the drive systems of drive units of bucket-chains and drives units of slewing motion which are the main sources of noise.

As the research objects, they were two similar 30-odd years old bucket-chain excavators (Fig. 2). The first of these machines built-in 1960 was not modernized before investigations. The second tested excavator built-in 1956 has been subjected to rebuilding while the old main drive units have been replaced with new ones.

The noise in the zones of the main drive units, i.e., the bucket-chain drive unit and the rotational drive unit (Fig. 3) was investigated. The bucket-chains in the tested excavators were powered by drive units of 200 kW whereas the slewing motion was driven by unit of 14 kW (in not modernized excavator) and by unit of 15 kW (in modernized excavator).

Exemplary acoustic spectra in the zones of main drive units of both long-term operated chain excavators are shown in Figs. 4 and 5.

The research results have shown that there were significant acoustic hazards in the zones of main drive units of the not modernized bucket-chain excavator. The total noise level was around 90–96 dB(A). Under these conditions, the permissible exposure time to noise in the engine room without additional hearing protection (earmuffs) was 3–20 min. Such a short period significantly limited the possibility of making ongoing inspections during the normal operation of this excavator.

For the purpose of comparison, noise tests of the main drive units of the second modernized bucket-chain excavator were also carried out. This modernization con-

Fig. 2 Research object—bucket-chain excavator

Fig. 3 Drive unit of bucket chain (on the left) and drive unit of slewing motion (on the right) of the tested bucket-chain excavator

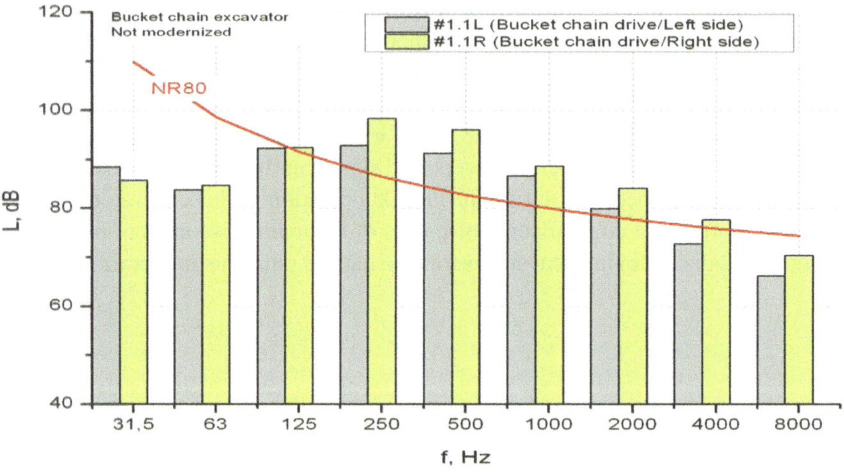

Fig. 4 Typical noise spectrum of the bucket-chain drive unit of a long-term operated excavator (not modernized)

sisted of, among others the implementation of a new bucket-chain drive unit powered by two hydraulic motors.

The reduction of acoustic activity of main drive units was one of the principal aims of modernization of this long-lasting bucket-chain excavator. A comparative analysis of noise research of both excavators showed unambiguously that this goal was achieved because a quite significant reduction of noise in the zone of main drive units was obtained (Fig. 6).

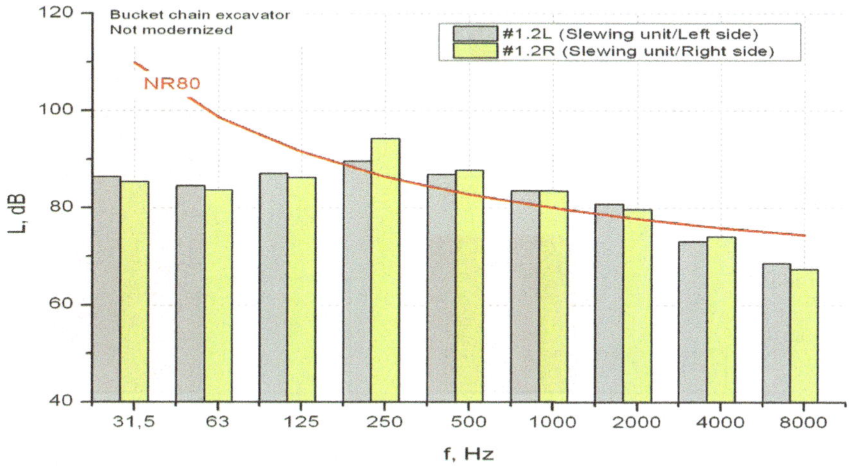

Fig. 5 Typical noise spectrum of the slewing drive unit of a long-term operated bucket-chain excavator (not modernized)

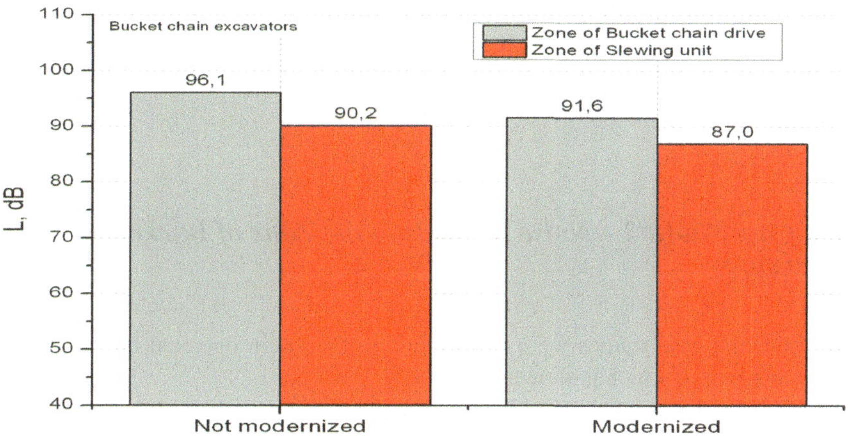

Fig. 6 Impact of modernization of bucket-chain excavators on the overall noise level in the main drive unit zones (the upper limits of 95% confidence intervals on mean values are shown)

As a result of the modernization of drive units of chain excavators, the noise level has been reduced as follows (Fig. 6):

- by about 6 dB in the zone of bucket-chain drive,
- by about 4 dB in the zone of drive unit of slewing motion.

The theoretical values of permissible exposure time for noise t_{exp} in the zones of main drive units of both tested bucket-chain excavators was estimated as follow (see Fig. 7):

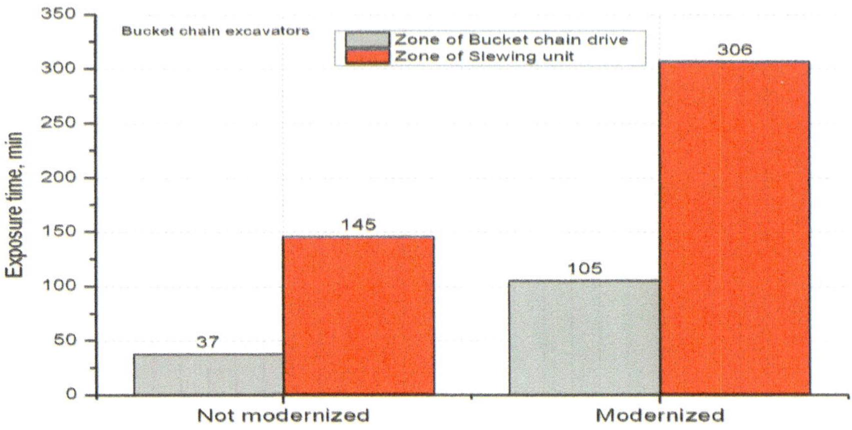

Fig. 7 Impact of modernization of bucket-chain excavators on the permissible exposure time to noise in the zones of main drive units

- $t_{exp} = 37–145$ min in the long-term operated excavator before modernization,
- $t_{exp} = 105–306$ min in the long-term operated excavator after modernization.

Therefore, the results of the authors' research have shown that this goal of modernization of long-term chain excavators, i.e., considerable noise reduction has been achieved.

5.2 *Case Study#2—Noise Hazards in the Zone of Bucket Wheel*

The following case studies were carried out on long-term operated bucket-wheel excavators, one of which is shown in Fig. 8.

The first of these cases concerned the assessment acoustic hazards in the zone of bucket-wheel drive and in the local discharging zone (Fig. 9).

The assessment of the noise level in these zones is particularly important due to the fact that the operator's cabin is located quite close to the bucket wheel.

The results of noise tests are shown in Figs. 10 and 11. It has been proved that in the bucket-wheel zone the average overall noise level exceeds the limit values $L_A =$ 85 dB(A) only on the left side of the bucket wheel (where the power unit is installed). The root mean square values of noise (upper limits of 95% confidence intervals) in these zones were the following:

- $L_{A(upper)} = 91$ dB(A) on the left side of bucket wheel,
- $L_{A(upper)} = 83$ dB(A) in the central zone on a service platform above the bucket wheel,
- $L_{A(upper)} = 85$ dB(A) on the right side of bucket wheel.

Fig. 8 Research object—bucket-wheel excavator

Fig. 9 Bucket wheel: drive unit zone (on the left) and local discharging zone (on the right)

Under these conditions, the permissible exposure time to noise in the zone of bucket wheel, estimated on the basis of the maximum total value $L_A = 91$ dB(A), is about $t_{exp} = 120$ min.

However, it should be borne in mind that when mining the overburden with rock inclusions, the noise was larger and extreme PEAK values reach the level over 100 dB (Figs. 10 and 12). This caused significant discomfort in the bucket-wheel zone.

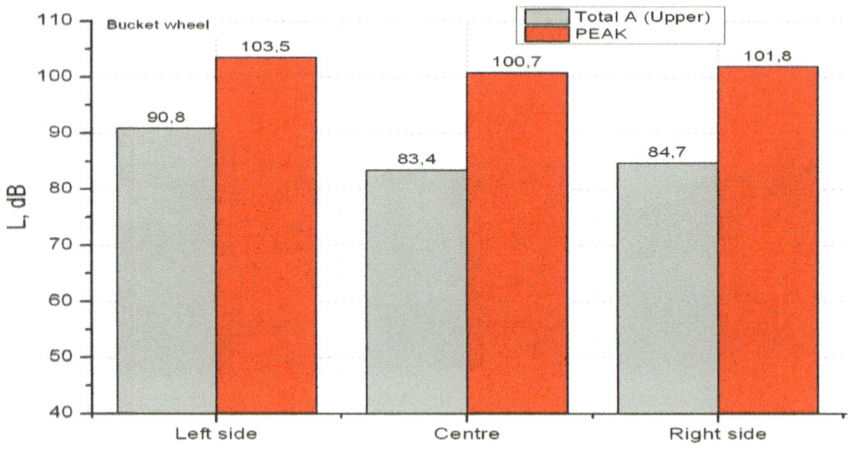

Fig. 10 Total noise level L_A (gray bars) and PEAK values of noise (red bars) in the zones of bucket wheel of a excavator (for the L_A level the upper limits of 95% confidence intervals on mean values are shown)

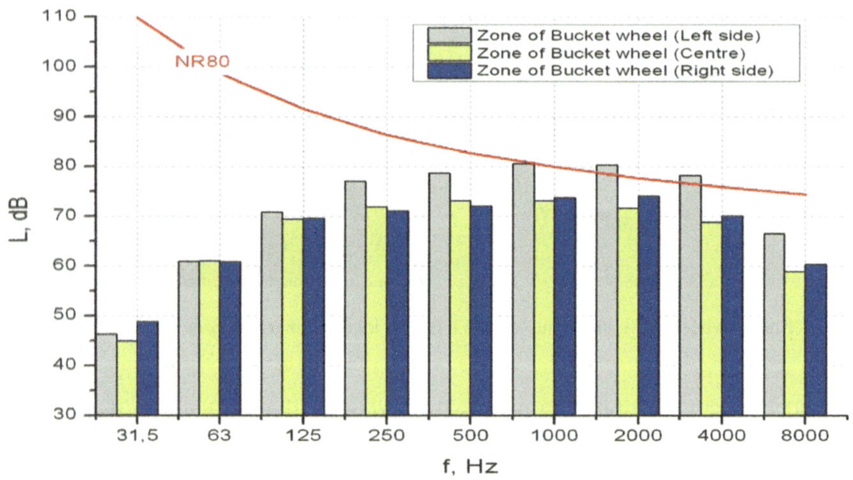

Fig. 11 Typical noise spectrum in the zones of bucket-wheel during removing the overburden without stones

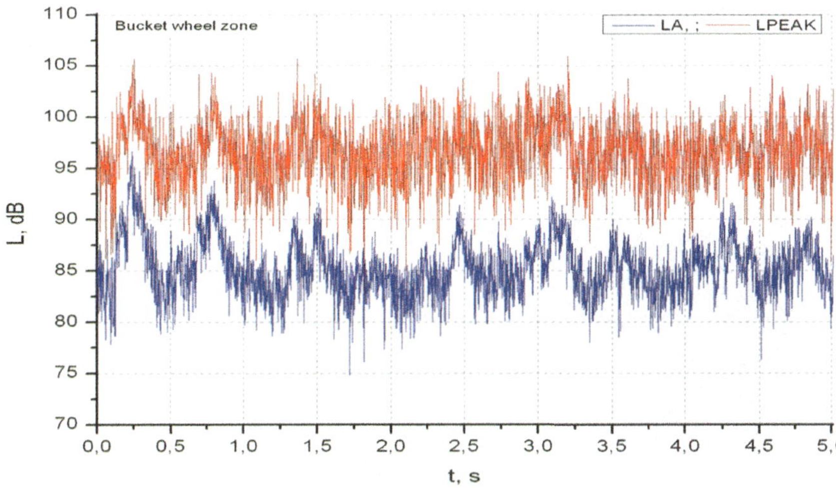

Fig. 12 Typical time history of noise level: RMS values (blue line) and PEAK values (red line) in the bucket-wheel zone

5.3 *Case Study#3—Noise Hazards in the Central Discharging Zone and in the Zone of Drive Unit of Belt Conveyor in a Bucket-Wheel Excavator*

As a next case study, the noise assessments in the central discharging zone (Fig. 13) and in the zone of drive unit of main band conveyor (Fig. 14) in a bucket-wheel excavator were carried out.

The main results of noise research in the zones of central discharging and drive unit of belt discharging conveyor are presented in Fig. 15.

The tests have shown that there were noise hazards in these zones. The total level and peak values of noise and the permissible exposure time were as follows:

- in the central discharging zone:
  - total values $L_{A(upper)} = 98$ dB(A), peak values $L_{PEAK} = 108$ dB and permissible exposure time for noise $t_{exp} = 24$ min.
- in the zone of drive unit of a main-belt conveyor:
  - total values $L_{A(upper)} = 101$ dB(A), peak values $L_{PEAK} = 98$–109 dB and permissible exposure time for noise $t_{exp} = 12$ min.

Examples of noise octave spectra in these zones are shown in Fig. 16. The analysis of the research results has shown that exceeding the value of the relevant Noise Rating Curve NR80 occurs in the spectral bands $f = 500$–2000 Hz.

Fig. 13 Central discharging zone in a bucket-wheel excavator

Fig. 14 Zone of drive unit of the main band conveyor in a bucket-wheel excavator

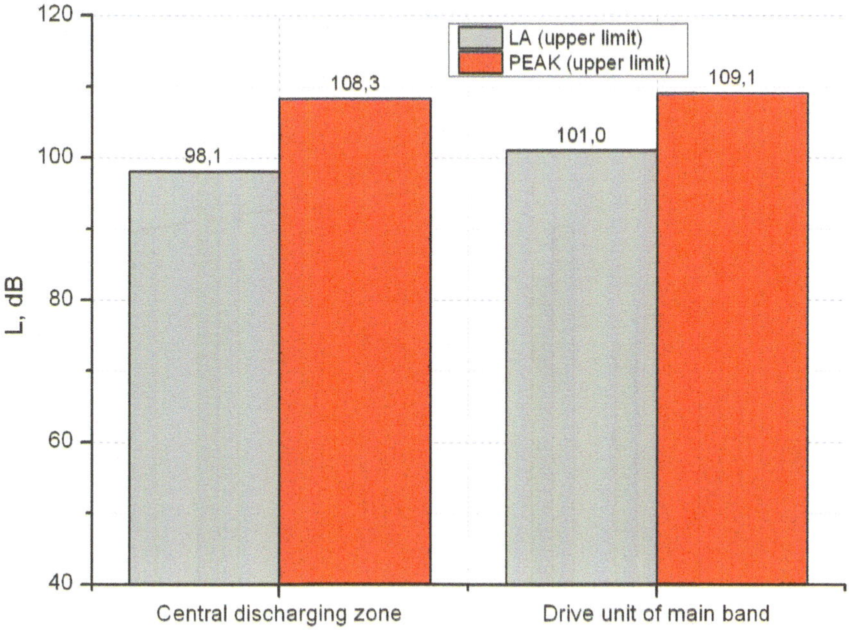

Fig. 15 Total noise level L_A (gray bars) and PEAK values of noise (red bars): in the discharging zone and in the zone of drive unit of band conveyor in a bucket-wheel excavator (for the L_A level the upper limits of 95% confidence intervals on mean values are shown)

5.4 *Case Study#4—The Noise Hazards in the Zones of Drive Units of Slewing Motion of a Bucket-Wheel Excavator*

As a next case study the noise assessment in the zones of slewing units of a bucket-wheel excavator was carried out. The tested mechanism was equipped with 3 drive units of 55 kW each. One of these drive units is presented in Fig. 17.

The relevant noise research was carried out during continuous removing of overburden when the superstructure was turned alternately clockwise (right turn) and anticlockwise (left turn). The research results are presented in Figs. 18 and 19.

The research has shown that there were some not high noise hazards in the zones of drive units of slewing motion of a tested bucket-wheel excavator (Figs. 18 and 19). The greatest values of hazards occurred in the drive unit#3 where the total noise level was $L_A = 87$–91 dB(A). Under these conditions, the permissible exposure time to noise has been estimated as $t_{exp} = 145$ min.

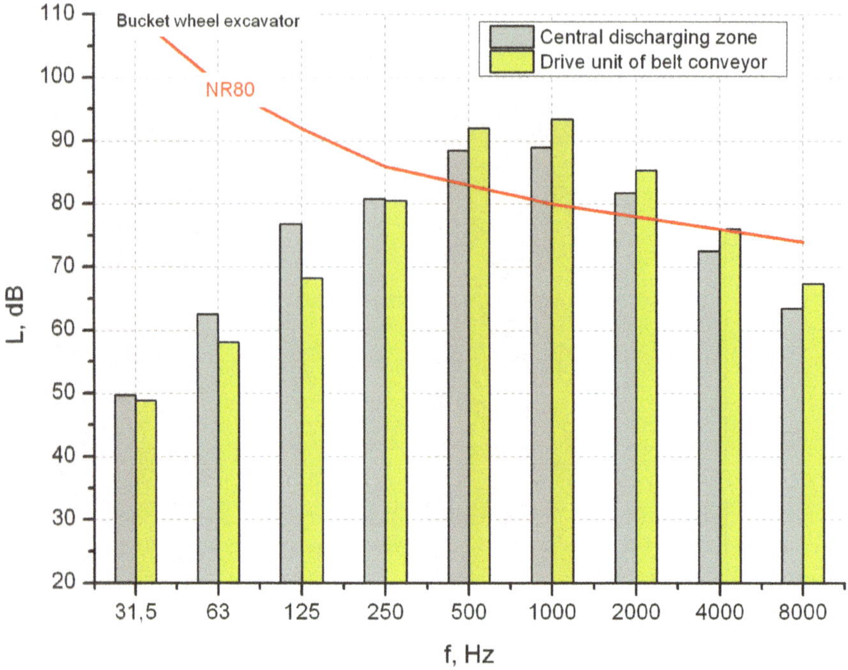

Fig. 16 Typical noise spectra: in the central discharging zone (gray bars) and in the zone of drive unit of belt conveyor (yellow bars) in a bucket-wheel excavator

Fig. 17 Zone of drive unit of slewing motion of a bucket-wheel excavator

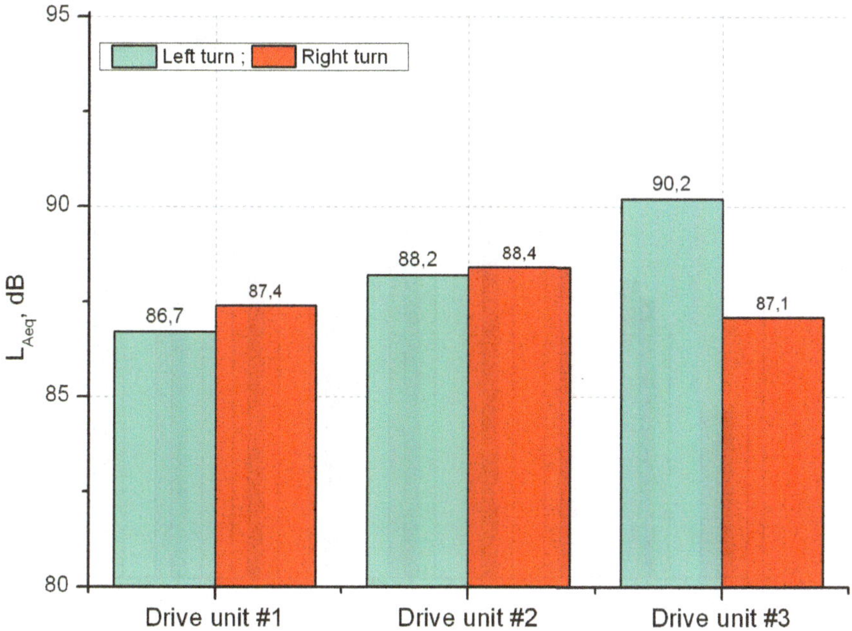

Fig. 18 Total noise level in the zones of the drive units of slewing motion of a bucket-wheel excavator (the upper limits of 95% confidence intervals on mean values are presented)

5.5 Case Study#5—The Noise Hazards in the Zones of Drive Units of Belt Conveyors in a Large Size Stacker

The last presented case concerns the noise hazards in the zones of drive units of main-belt conveyors in a large size long time operated stackers.

As a research object, the largest stacker in Polish brown coal mines was selected (Fig. 20). The theoretical capacity of stacking was $Q_t = 15{,}400\ m^3/h$, and the stacking radius $L = 120$ m. At the time of testing, the machine was over 30 years old.

The research was carried out in the zones of drive units of main-belt conveyors, i.e., the loading conveyor and the stacking conveyor. Each belt was powered by a twin-engined unit with an installed power of 2 × 1120 kW. Some exemplary research results are shown in Figs. 21, 22 and 23.

The research has shown that noise in the zones of drive units of main conveyors of a tested stacker significantly exceeded the level of 85 dB(A) permissible for a nominal

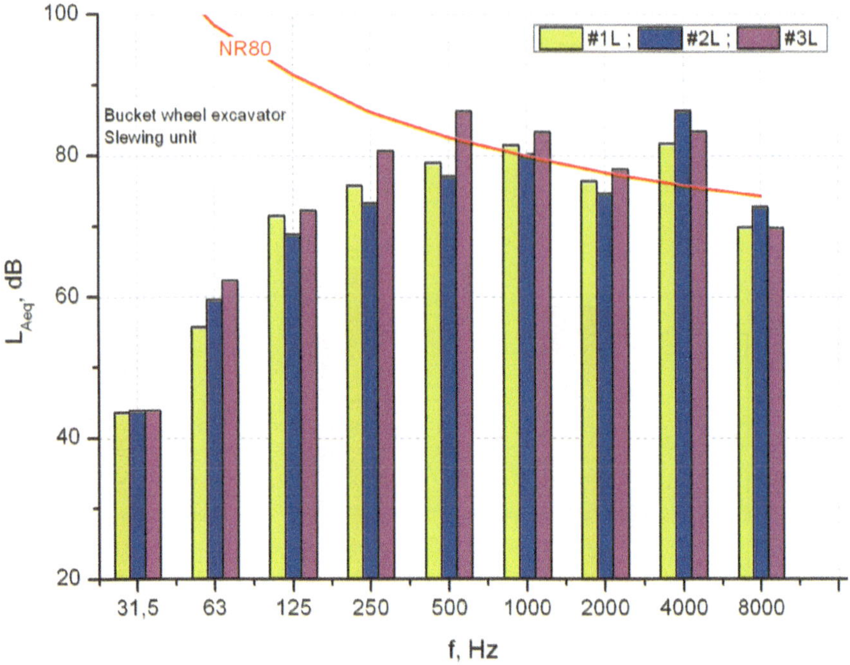

Fig. 19 Typical noise spectrum of the slewing drive units of a bucket-wheel excavator

Fig. 20 Research object—a long-term operated large-size stacker

eight hour exposure period. The equivalent total noise level reached values of $L_{A(max)}$ = 105–108 dB(A). The largest exceeding of the relevant Noise Rating Curve NR80 occurred in the band f = 125–4000 Hz and amounted up to 19–22 dB(A).

In these conditions, the permissible exposure time to noise in the zones of main drive units was assumed as follows:

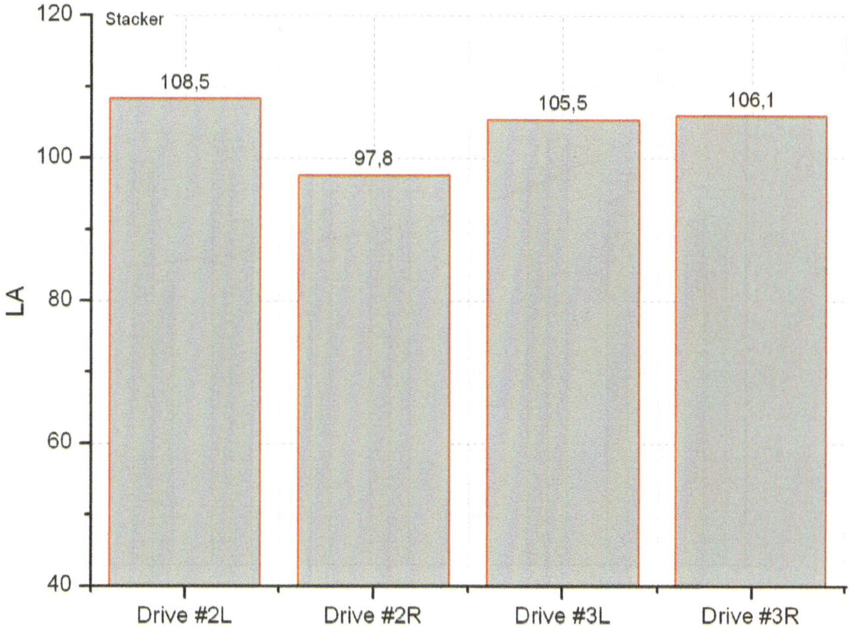

Fig. 21 Total noise level in the zones of drive units of loading conveyor (drive #2L on left side, drive #2R on right side) and stacking conveyor (drive #3L on left side, drive #3R on right side) of a large-size stacker (the upper limits of 95% confidence intervals on mean values are presented)

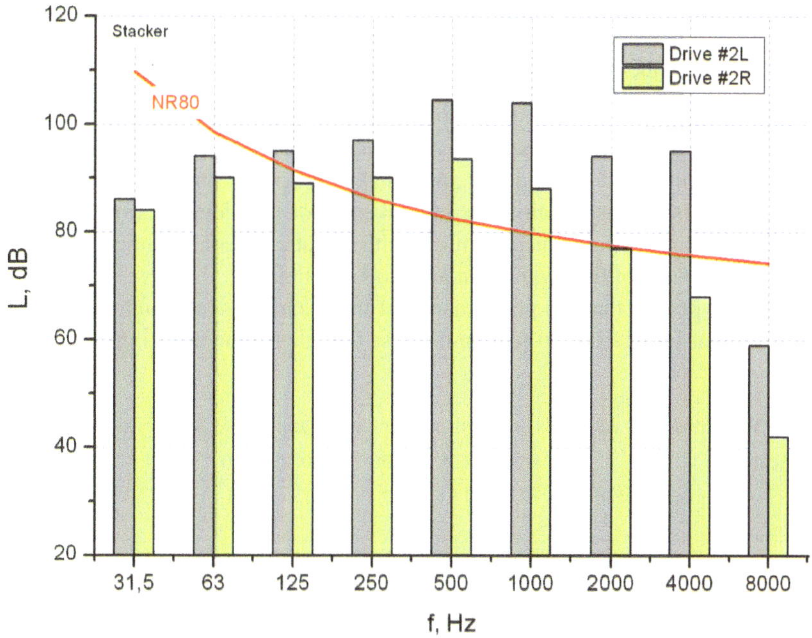

Fig. 22 Typical noise spectrum in the zone of drive units of loading conveyor in a large-size stacker (drive #2L—on left side, drive #2R—on right side)

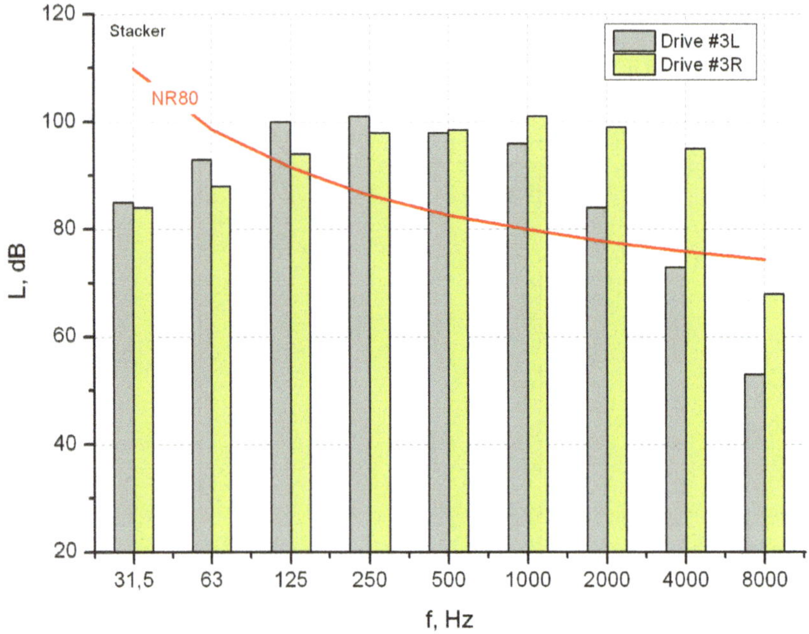

Fig. 23 Typical noise spectrum in the zone of drive units of stacking conveyor in a large-size stacker (drive #3L—on left side, drive #3R—on right side)

- $t_{\text{exp}} = 2$ min in the zone of drive units of loading conveyor,
- $t_{\text{exp}} = 4$ min in the zone of drive units of stacking conveyor.

Such a short permissible exposure time to noise made practically impossible to perform ongoing technical inspections of the stacker drive units and some means of hearing protection must be used (e.g., personal hearing protection devices).

Taking this into consideration, the effectiveness of individual hearing protectors was evaluated, assuming some ear muffs with the attenuation parameters given in Table 1. The results of this evaluation are shown in Fig. 24.

Thus, it is proved that the performance of these earplugs taken into account are quite sufficient for hearing protection because the relevant noise rating curve NR80

Table 1 Performance of ear muffs assumed as a personal hearing protection device

Frequency (Hz)	63	125	250	500	1000	2000	4000	8000
Reduction of noise level (dB)	12.8	11.6	17.2	21.7	30.4	29.2	35.4	24.4

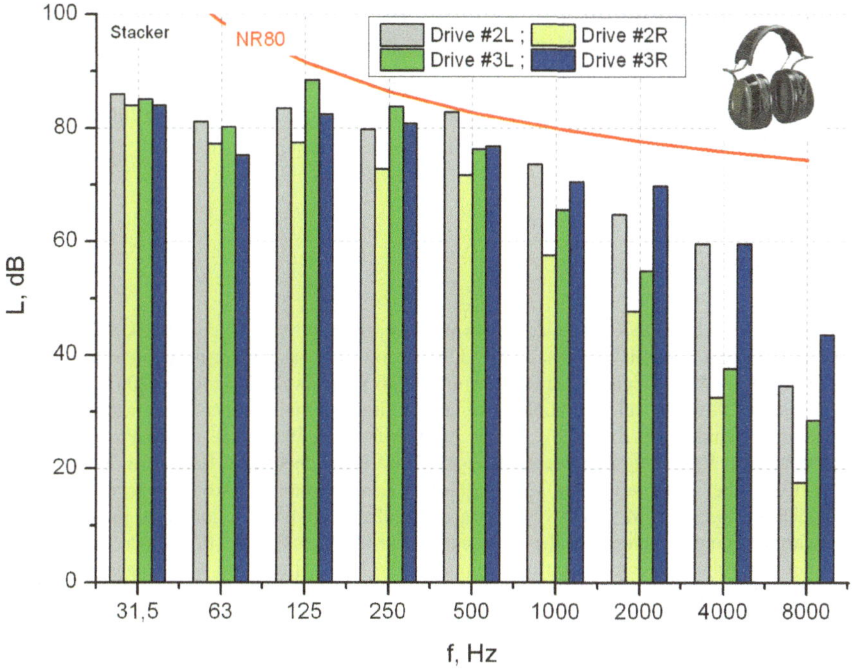

Fig. 24 Noise spectrum in the zones of main drive units [(drive #2 and drive #3) in the tested stacker (drive #3L—on left side, drive #3R—on right side] in the simulated conditions of the use of earmuffs

is not exceeded by the noise spectra (s. Fig. 24). Then periodical inspections of main drive units in the tested stacker may be carried out without restrictions.

6 Summary

Increase in noise hazards is one of the important attributes of degradation processes in long time operated basic machines for open-pit mining.

The research carried out in all the Polish brown coal mines has proved that the noise hazards are the essential problem and should be not underestimated in any way.

The presented examples are typical for all the large-size excavators and stackers. One of the effective ways to solve these problems of noise hazards is to reduce the acoustic intensity of noise sources. This conclusion applies in particular to the drive units. In addition, some personal hearing protection devices (e.g., ear muffs) can be used as a temporary precaution.

Since the mid-1990s, a comprehensive modernization program for basic machinery has been consistently implemented in the Polish brown coal industry. As a part of this program, new driving systems are implemented, among other things.

Research carried out by the authors of this chapter as well as by other team members proves the high effectiveness of modernization activities in the context of reducing the level of acoustic hazards in the large size machines for open-pit mining.

Additional methods of lowering noise and vibration level in different machines are used. Some examples are presented in [6–8].

References

1. Babiarz, S., Dudek, D.: Kronika katastrof i awarii maszyn podstawowych w polskim gornictwie odkrywkowym [Chronicle of disasters and break-downs of basic machinery in Polish open pit mining]. Edited by Oficyna Wydawnicza Politechniki Wroclawskiej, Wroclaw (2007) (in Polish)
2. Dudek, D., Nowakowski, T.: Problems of degradation and maintenance of surface mine engineering machines. In: Bicego, V., Nitta, A., Viswanathan Warley, R. (Eds.) Proceedings of International Symposium on Materials Ageing and Component Life Extension, Milan, Italy, 10–13 October 1995, Vol. 2, pp. 1285–1294. Engineering Materials Advisory Services
3. Dudek, K., Sokolski, M.: Improved vibroacoustic characteristics—A goal to be pursued in the open pit mining machinery of the XXI century. In: Golosinski, T.S. (Ed.) Proceedings of the American-Polish Mining Symposium Mining in the New Millennium. Challenges and opportunities, Las Vegas, Nevada, USA, 8 October 2000, pp. 193–199. Brookfield, A. A. Balkema, Rotterdam
4. Sokolski, M.: Wysoki poziom zagrozen wibroakustycznych – atrybut degradacji maszyn podstawowych gornictwa odkrywkowego [High vibroacoustic hazards level - attribute of the degradation of the open pit mining machinery]. Gornictwo Odkrywkowe, R. 49, nr 3/4, pp. 148–152 (2007) (in Polish)
5. Sokolski, M., Sokolski, P.: Acoustic climate in the cabins as a factor of rebuilding effectiveness of long term operated bucket wheel excavators—a case study. In: 23rd International Conference Engineering Mechanics 2017, Svratka, Czech Republic, 15–18 May 2017, pp. 906–909
6. Duvigneau, F., Koch, S., Woschke, E., Gabbert, U.: An effective vibration reduction concept for automotive applications based on granular-filled cavities. J. Vib. Control **24**(1), 73–82 (2016)
7. Fiebig, W., Wrobel, J.: System approach in noise reduction in fluid power units. In: Proceedings of the ASME-BATH Symposium on Fluid Power and Motion Control, V001T01A025 (2018)
8. Polivaev, O.I., Kuznetsov, A.N., Larionov, A.N., Beliansky, R.G.: Reduction of external noise of mobile energy facilities by using active noise control system in muffler. In: IOP Conference Series: Materials Science and Engineering, pp. 327 (2018)

Problems of Research and Operational Assessment of Earth-Moving Machines

Marek Młyńczak

Abstract Earth-moving machine is operated in various conditions of use and maintenance, which results in varying intensity of aging and degradation. It also hinders rational management of operation aiming at the most efficient use of the object. A key element of this efficiency is getting the highest effect in relation to investment. It is proposed a method of machines profitability assessment with limited historical operating data. The assessment was based on interim data from the use of machinery. Functions which characterize the period of the rational working period of machines are developed. The examples of evaluation and variability of proposed characteristics for different types of machines are shown.

Keywords Earth-moving machine operation · Operational test · Maintenance · Reliability

1 States and Variability of Operation Process

Classical assessment of technical object reliability assumes two states of availability indicating the ability to perform designed functions. In both states, it is also assumed that the object is fully subjected to processes assigned to these states, i.e.,in usage state, it fulfils given tasks (work) and during maintenance, it is restoring to full availability.

Operation reality is, however, richer and is much more intermediate states in relation to the shown state. Figure 1 shows the decomposition of operational states recognized in accordance with the standard recommendations [1], and similar classification due to availability states is shown in Fig. 2.

Shown decomposition of states indicates the possibility of a diverse approach to the assessment of the object in operation, as well as assignment and relationship among reliability and operational states.

M. Młyńczak (✉)
Faculty of Mechanical Engineering, Wrocław University of Science and Technology, ul. Wybrzeże Wyspiańskiego 27, 50–370 Wrocław, Poland
e-mail: marek.mlynczak@pwr.edu.pl

M. Sokolski (ed.), *Mining Machines and Earth-Moving Equipment*,
https://doi.org/10.1007/978-3-030-25478-0_5

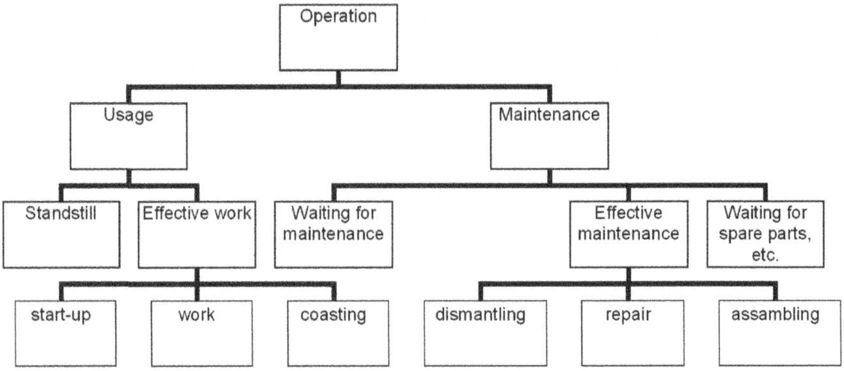

Fig. 1 Decomposition of operational states (example)

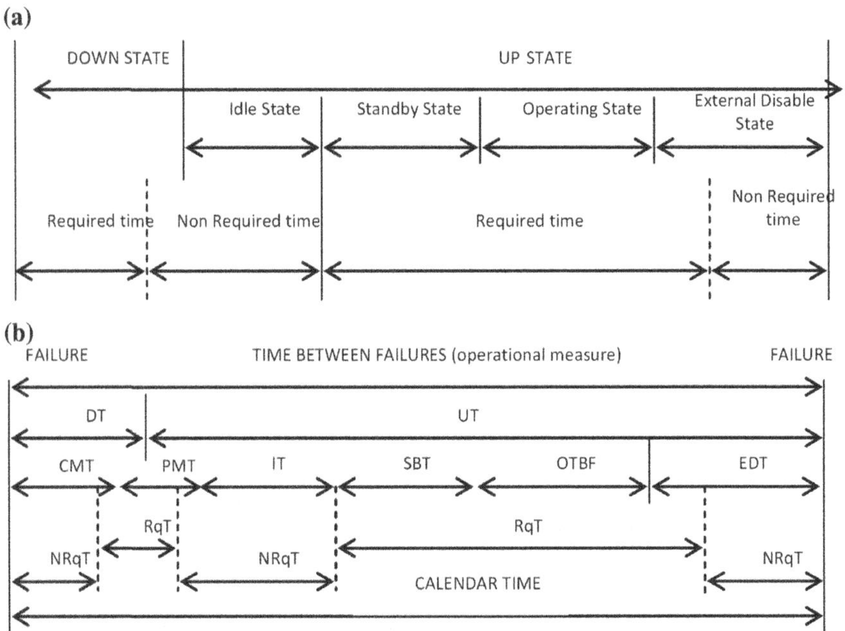

TBF	time between failures	SBT	standby time
DT	down time	OTBF	operating time between failures
UT	up time	EDT	external disabled time
CMT	corrective maintenance time	RqT	required time
PMT	preventive maintenance time	NRqT	non-required time
IT	idle time		

Fig. 2 Decomposition of availability states relating to operational states; based on dependability and quality of service—IEC 60050–191

Operation management has to balance between profit and operating expenses. Looking for high-profit manager tends to shorten downtime by limiting number of prophylactic maintenance but it may result in higher failure intensity and unexpected breakdown costs (Fig. 3). Knowledge about reliability and planned tasks may define reasonable preventive measures with minimal risk of mission failure.

Optimal period of preventive maintenance looks simple, but it requires probabilistic parameters of the operation process [2, 3]. Analytical solution of the problem is complex because of the complexity of processing distribution function describing random processes [2].

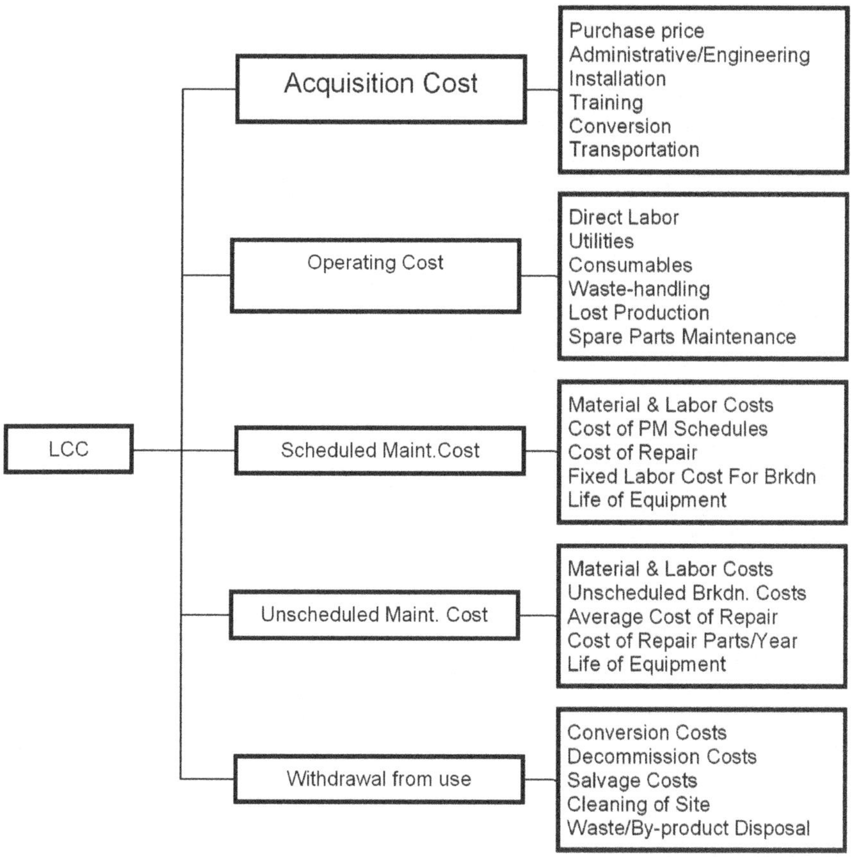

Fig. 3 Operational cost classification

2 Measures of Operational Effectiveness

2.1 Economic Assessment of Operation

Cost models are shown as Life Cycle Cost (LCC) [4–6] take into consideration many operational costs grouped into four categories (Fig. 2). Main costs are related to the state of the object: prior to usage (object acquisition), usage (true making use), scheduled maintenance, repairing and withdrawal from use. Acquisition costs cover important costs before an object is installed and not only a price but also administration, staff training, preparation of the infrastructure and transportation to the end-user. Operation consumes money for energy, labor, wastes handling and also may contain losses of unattainable profit. Maintenance costs are related mainly to diagnostics and repairs with spare part, operational materials, labor cost, and technical infrastructure. Final withdrawal from use costs covers decommissioning and costs due to environment protection [7–9].

Good practice in machine operation encourages making frequent predictive maintenances counteracting random failures. Unfortunately, this strategy increases costs as more prevention is done and stops an object from time to time. On the other hand, breakdown costs decrease as more predictive maintenance is performed. As user incurs both costs, he has to optimize the maintenance strategy keeping total cost at a minimum by reasonable predictive maintenance frequency (Fig. 4).

Corrective maintenance costs rise monotonically due to ageing, wearing and degradation processes. Example of corrective maintenance cost function for copper mine loader is shown in Fig. 5. It is strongly correlated with hazard rate function and is visible period of constant hazard rate period (1–36 months) and aging period (37 months to the end of loader life).

Treating the operation as part of a phase of object life cycle, costs of object development (design, tests, manufacturing, marketing) should be considered as constant, stable, which operation management generally does not have any impact. Models of operating costs are classified according to various criteria, such as type of costs, object/place of generating cost, cost carrier, cost variability [10], and new

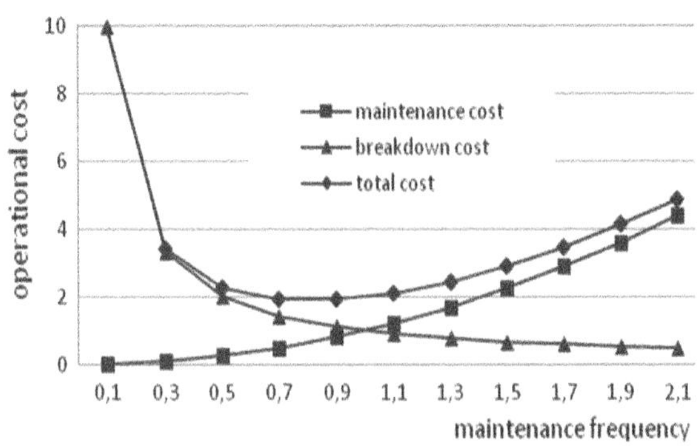

Fig. 4 Operational costs versus maintenance and failure costs

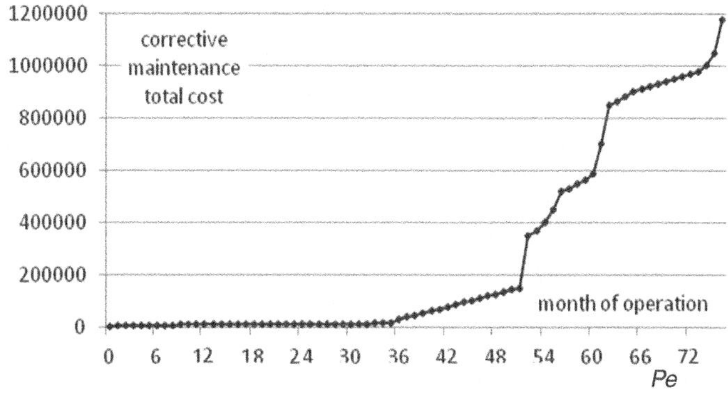

Fig. 5 Cumulative corrective maintenance cost in copper mine loader

management approach is to analyze the costs by operation phase known as ABC (Activity-Based Costing) [11–13]. The idea behind ABC is to assign costs to activities (activities) and resources and then to the final products or services (activities require resources and products are made from the resources used). According to this idea costs identification requires:

- identification and classification of activities,
- assignment of necessary resources costs to activities,
- identification of resource costs and determining the distribution of costs to action,
- determine the share of activities in the products/services,
- assignment costs to the cost of products/services.

Efficiency of ABC method is seen in saving time, improving quality, lowering costs and observation of trends [11].

The analysis of the operations shows the three main groups of activities related to machine use, maintenance and its management [2, 10, 14], wherein each of these groups can be distinguished specific logistics operations (support).

The actions correspond to the states in which resource stay, hence the states of decomposition leads to the identification of actions and assigned costs. Insight decomposition is limited, in turn, a resolution of the perception of operational situation, which results in reducing the number of data or the necessity of their gathering, storing and processing. Determination of dependence on test costs from decomposition level of the system is subject to considerable uncertainty, which is limited using statistical methods, Bayesian theory and based on fuzzy sets [15, 16].

Cost analysis based on the main processes observed in operation shows the four main groups of costs: administrative costs, operation, maintenance and additional costs so-called external costs arising from the effects of undesired events (1).

$$OC = AC + UC + MC + EC \tag{1}$$

where

OC	operational costs,
AC	administrative costs; AC = IC + MMC,
IC	costs of infrastructure maintaining (constant, infrastructural),
MMC	costs of operation and risk management,
UC	usage costs, UC = UPC + UIC + UAC,
UPC	costs of passive usage (standstill, standby),
UIC	cost of idle state,
UAC	cost of active work,
MC	maintenance cost, MC = MCP + MCPR + MCCR,
MCP	costs of passive maintenance (waiting for maintenance),
MCPR	costs of active predictive maintenance,
MCCR	costs of active corrective maintenance (repair),
EC	external costs of undesired events (losses, accident damages).

The specified costs are dependent on many variables characterizing the operational process, such as the number of technical objects, age and quality of these objects, size of the crew, a measure of amount of work done, the quantity and quality of material resources, the size of operational database, financial resources and technological skills and psychophysical abilities of people, systems, logistics support, environmental conditions, volatility of the economic situation, etc. Estimating the effectiveness of operational test in terms of operating costs as a function of these variables is only possible by strong simplifying assumptions and obtaining mostly confidential economic information.

Importantaspect of operational economics is amortization defined as the loss of value of fixed assets caused by its degradation and economic aging resulting from technological progress associated with the possibility of obtaining on the market, e.g.,modern machines, more efficient, cheaper in operation, giving higher quality products [17]. Depreciation allowance creates funds for the renewal of used fixed asset. In amortization assessment there are two measures exist:

- percentage ratio fully amortized machines in machine population:

$$A_{\%} = \frac{M}{N} \cdot 100\% \tag{2}$$

- average level of amortization in machine population:

$$A_{\mathrm{av}} = \frac{1}{N}\sum_{i=1}^{N}\left(\frac{C_{Ai}}{C_{Zi}}\right) \cdot 100\% \tag{3}$$

where

M number of fully amortized machines,
N number of all machines,
C_{Ai} total depreciation for ith machine,
C_{Zi} purchase cost of ith machine

In the group of 46 wheel loaders, only one machine was not fully amortized what indicates that fleet of earth-moving machines is old and management should consider exchange of the oldest objects.

2.2 Operational Effectiveness Assessment

An important aspect of operation is a choice of proper operational strategy, which is based on several criteria, among which the most important are: cost, quality, flexibility, speed, and reliability [6, 18, 19]. These criteria should be completed by safety in a wide range of effects [3, 20]. These criteria are the basis for decision-making in all areas of the engineering, material and human resources.

The most important strategies that capture the above criteria are [18]:

- strategy by the volume of production/services (adjust the size of the operation effect to the needs and material resources),
- strategy for the use of human resources (training, remuneration, motivation, management),
- strategy according to the quality of production/services (objectives, measures, assessment of the opportunities, benefits),
- strategy according to the sources of production/services (proportion on own production and outsourcing, selection and evaluation of suppliers),
- strategy by using activity resources (scope of the system operation to meet and respond to the decisions).

The strategy of system resource utilization can be considered more detailed similarly to management strategy according to [21]:

- reliability of mechanical objects consisting of providing the highest possible rate of system availability,
- economic efficiency minimizing operational costs of mechanical system,
- available technical resources servicing according to the amount of work done,
- system technical condition allowing the most efficient use of technical resources.

The choice of strategy is a complex task in technical, economic and organization domain that requires knowledge of both the market needs (end-user of the product or services) and the total balance of costs and profits and technical effectiveness of operated devices. Evaluation of selected strategies should be based primarily on the obtained result of the system as intended in design and the need of the user. Such assessment is based on the assessment of the overall effects of the system,

OEE (Overall Equipment Effectiveness) [22] which takes into account availability, efficiency and performance (4).

$$\text{OEE} = \text{AVAILABILITY} \cdot \text{PERFORMANCE} \cdot \text{QUALITY} \tag{4}$$

where

AVAILABILITY	Availability index (ratio of upstate in planned operational time),
PERFORMANCE	Efficiency index (ratio of active usage time in upstate time),
QUALITY	Quality index (ratio of good products/services in total number of products)

It is estimated that the expected (the world average) value of OEE is, appropriately taking into account the given factors of the product, as [22].

Considering the given approach, supplemented by safety aspect, one can evaluate the effectiveness of the system taking into account also compensation for losses due to unexpected failures or undesired events caused by third party [23] (5).

$$TEE = A \cdot Pe \cdot Q \cdot S \tag{5}$$

where

A availability index,
Pe performance index,
Q quality index,

$$S = 1 - \frac{CL}{PR}$$

S safety index,
CL cost of losses, damages $[0; PR]$,
PR gained profit

Quality index is a measure of system resilience, which is understood as resistance of the system to disturbances and losses. If the system is provided by safety measures it is more failure, human error, and environment influence-proof and suffers less in case of undesired event appearance. Proper applied safety measures, according to BATNEC (Best Available Technology Not Entailing Excessive Cost) giving the highest returns with the lowest possible expenditure on security [24]. Losses in the magnitude of catastrophe should be included in the formula above to the amount of income. Events having such large consequences are unacceptable and in principle cause the liquidation of the enterprise.

3 Usability Assessment of Earth-Moving Machines

In many technical systems machines and devices with similar functionality are operated, but with different ages, different levels of wear and operational potential or from different manufacturers. Observed is also uneven use of such machines due to changing tasks, and even operator preferences. Appearing periodically in the operation problem of determining threshold of economic viability of use of appliances is usually solved on the basis of mileage (loss of operational potential, amount of work done), or based on the age of the device.

Observed is a convergence of expert opinions that the old machines with the physical and so-called moral warring are fully depreciated and also have high values of total runs.

The decision about withdrawal the object from operation is a complex problem that requires taking into account aspects such as current cost of its operation, its availability, security, or external demands such as for instance manufacturer recommendation due to the period of safe and efficient operation.

Examples of strategies replacement devices are based on:

- replacing the unit after a period of depreciation,
- replacing the unit after the period recommended by the manufacturer or other users,
- replacing the unit on the basis of reliability,
- replacement of equipment after reaching the limit of OEE [22],
- replacing the unit on the basis of complex assessment measures (e.g.,score assessment method).

3.1 Machine Replacement After a Period of Depreciation

The method of machine replacement after a period of depreciation consists of replacement with similar, new one, fulfilling the same function. It is easy to use a method that no knowledge is required about the history of machine operation. The advantage of this method is that the machine stock is renewed periodically maintaining a high level of modernity and availability.

The method does not take into account the actual degree of wear and operational potential and is completely contrary to the assumptions of the management operation on the basis of total cost (LCC).

3.2 Machine Replacement After Manufacturer or Other Users Recommendation

The method is based on the decisions of decision-maker none closely related to operational system, without the knowledge of the history and specifics of use and maintenance. Arbitrary decision about replacing the devices is not supported by observations from real process and it is not always rational.

3.3 Machine Replacement on the Ground of Dependability

Dependability of machine system is characterized using assessment measures, mainly in relation to reliability, availability, maintainability, and safety [1, 8]. These measures relate to three fundamental aspects of the operation: the ability for continuous operation (the fulfillment of tasks, missions), the ability to restore availability after failure, and the ability of the device to perform the function for a given level of confidence without damages.

The basic measure of reliability is a failure rate defined as a conditional probability of failure at a given moment of time, provided no damage to the moment. This function (Fig. 6) shows the relationship between main types of failures, and also allows determining reliability function (6):

$$R(t) = e^{-\int \lambda(\tau)\mathrm{d}\tau} \tag{6}$$

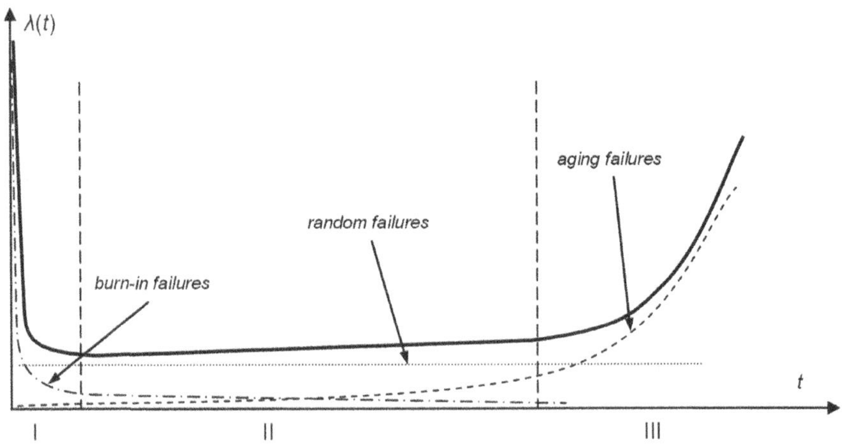

Fig. 6 General shape of hazard rate function (bath-tube curve)

where

$R(t)$ Reliability function,
$\lambda(t)$ Failure rate function (hazard rate function).

The shape of the failure rate function (Fig. 6) is typical for most of the technical objects mass-produced. For prevention issues, the most important part of this curve is a third period in which the function of intensity is increasing. Most defects observed in this period are ageing failures (fatigue, wear, corrosion). Monitoring relevant symptoms allow to some extent for the prediction of failure time.

The graph (Fig. 6) shows three phases of technical object life time:

I. Phase "burning-in" is characterized by failures caused by errors introduced pre-operation phase like design and/or production. It results also in errors in implementation, installation, and errors in adjusting the operating conditions for newly introduced to the operation of equipment as well as human errors came from lack of knowledge about machine operating. Hazard rate function is decreasing in that phase.
II. Phase of constant failure rate deals with failures arising from "normal". Observed failures are usually random, unexpected, difficult to predict caused by operators, environment or technical infrastructure in the operational neighborhood. Hazard rate function is constant in that phase.
III. Phase called as aging is the last phase of machine "life." Failures arising in this phase are the result of natural wear fatigue, corrosion, material degradation of the components of the object. Hazard rate function is increasing in that phase.

Anyway, it is observed the tendency to change the ratio of subsequent phases as higher quality control and engineering culture eliminates burn-in and ageing failures [7] and more failures have sudden, random nature.

If we consider a problem of determining age of machine replacement or withdrawal from use, it is important to take into account a criterion of efficiency. To ensure high availability of the system it is required to keep a slight excess of object quantity to achieve redundancy high and hence higher reliability In contrast, low operating costs can be achieved using medium quality devices with low energy consumption. An important criterion is also the maintainability, where both time to repair and cost of repair decide about machine usefulness.

The following reliability measures are most important while assessing the proper time of machine replacement (7–9):

- Availability:

$$A = \frac{UT}{T} \tag{7}$$

where

UT up-state time,
T operation time.

- Mean Time Between Failure:

$$MTBF = \frac{T}{n} \quad (8)$$

where

T operation time,
n number of failures (repairs) in the period of operation.

- Mean Time To Repair:

$$MTTR = \frac{RT}{n} \quad (9)$$

where

RT total repair time in the period of operation.

The disadvantage of this approach is the lack of taking into consideration the economic aspect of the operation and factors such as:

- moral aging understood as a lack of modernity of functional design, lower operator comfort, outdated way of working, causing the reluctance of employees to use the unit, etc.,
- lower productivity outdated equipment,
- difficult maintainability,
- difficulties in accessing supplies and spare parts.

Full description of random variable describing machine reliability is failure frequency and distribution functions. Example of empirical failure frequency function (histogram) and empirical distribution of a loader ŁK-5 is shown in Fig. 7. Figure 8 shows theoretical approximation of distribution function of time to failure of loader ŁK-2 [25] drawn in application Weibull++. Weibull distribution for time to failure is described by shape parameter $\alpha = 1.13$ and scale parameter $\beta = 39.15$.

3.4 Machine Replacement Based on Composition of Economic and Reliability Factors (CP Method)

Rational approach in the assessment of machine exchange period seems to be taking into account reliability and economic measures which depend both onusage and wear degree [26].

The basic characteristics, which were adopted to determine the limit of economic profitability of each use of the technological equipment is a function of the cost of operating the unit $C(t)$ representing an annual operating cost (10). The function is proposed depending on the age of the device t. The function includes elements

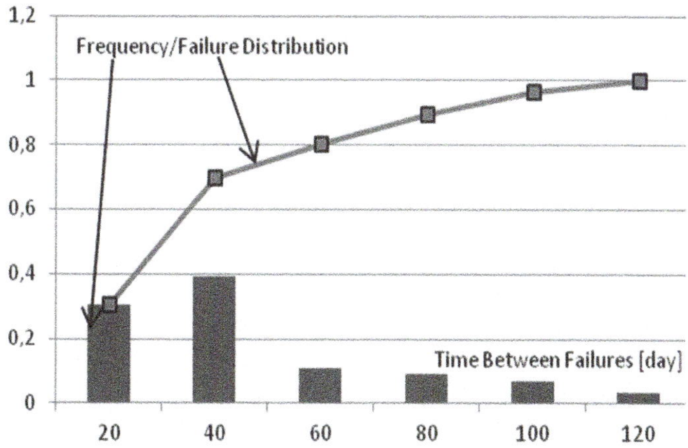

Fig. 7 Empirical failure frequency (histogram) and distribution functions of a loader ŁK-2

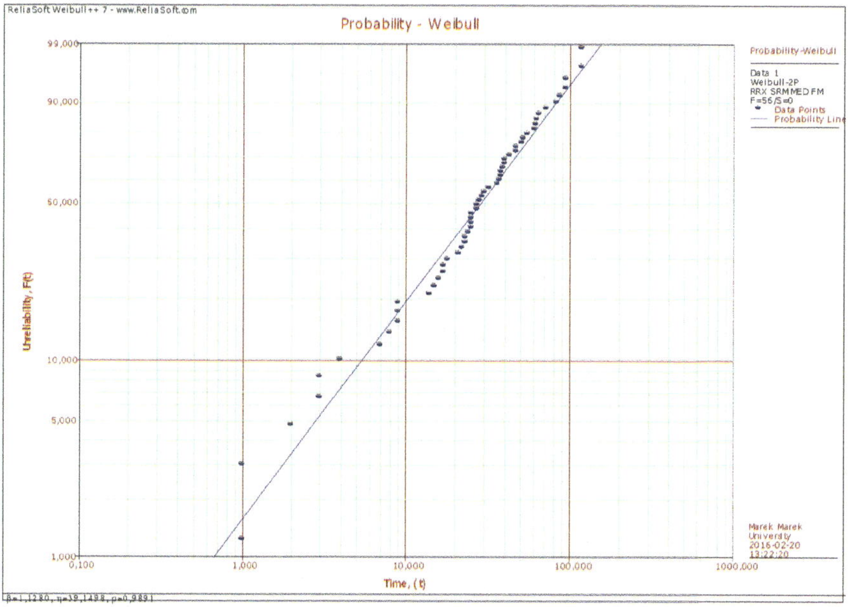

Fig. 8 Failure distribution approximation of a loader ŁK-5 using Weibull++ application

of economics such as the purchasing cost of the new machine, maintenance costs (materials and labor) and cost of lost income while repair and waiting for a job.

$$C(t) = \frac{C_M(t) + T_N(t) \cdot C_N + T_R(t) \cdot C_R + T_{RBH}(t) \cdot C_{RBH} + C_{NU}}{t} \tag{10}$$

where

$C_M(t)$ total cost of material and spare parts applied in maintenance until t,
$T_N(t)$ total time of maintenance until time t,
C_N unitary loss of income not gained while in maintenance,
$T_R(t)$ total time of availability until time t, but machine is idle,
C_R unitary loss of income not gained while idle,
$T_{RBH}(t)$ total number of man-hour spent in maintenance until time t,
C_{RBH} unitary cost of man-hour in maintenance,
C_{NU} purchase cost of new machine,
t mileage (time, moto-hour, etc.)

In the absence of sufficient information about the operation, evaluation is complemented by the function of profit $P(t)$ defining the surplus revenue derived by an object on the costs incurred on its maintenance and the cost of purchase, assuming that it is replaced at the age t (11).

$$P(t) = \frac{T_P(t) \cdot z - (C_M(t) + T_N(t) \cdot C_N + T_{RBH}(t) \cdot C_{RBH} + C_{NU})}{t} \tag{11}$$

where:

$T_P(t)$ total time of active usage until time t,
z unitary income

Functions $C(t)$ and $P(t)$ take into account both machine availability and their use and the cost of purchasing a new one incurred once but spread consequently over whole life. Analysis has shown that the increased costs of old machines are not higher than the cost of buying new, which of course is inefficient in terms of cost and availability. Theoretical approach requires determination of the appropriate extremes of function as a function of time corresponding to the best time for machine replacement. Process of machine aging should result in a decrease of availability and increase the cost of maintenances and uncompleted work.

4 Application of CP Method in the Study of Earth-Moving Machines [26]

4.1 Assumptions

The aim of the study was an attempt to estimate the replacement time using an approach based on functions $C(t)$ and $P(t)$ of a large set, of nearly 400, earth-moving machines. Machines have been grouped in 20 homogeneous groups where couple of them were loaders. One of the most difficult problem in calculations was met in data collection because average age of machines has reached 18 years and data required to calculate functions $C(t)$ and $P(t)$ were highly incomplete or even missing.

Due to the limited access to comprehensive data about the history of machines, research and analysis of the results were carried out with the adoption of appropriate assumptions for particular groups of machines and depending on the quantity and quality of the historical operating data. One of the major objectives for a particular machine that has not obtained any data about the operation was to assume that the older object has the same characteristics as the closest younger object. This procedure excluded unreal cases. This happened, for example, in relation to the machines aged 10–20 years, for which there were no data to support the costs incurred in normal use. During the data collection and analysis were also encountered numerous problems related to the periodic lack of data, its incompleteness, and written errors. This required verification, filtration, homogenization (e.g.,Time units) and interpolation for periods having no data.

Selected results are shown in graphical form. Set of results for each group of machines includes: availability function $A(t)$, use rate of the machine $U(t)$, intensity of machine use $IU(t)$ depending on its age and value of the objective $C(t)$ and $P(t)$ shown for the age range of 0–35 years.

Use rate expresses rate of upstate in total time: $U(t) = T_p(t)/t$, and intensity of use is yearly time of being in upstate in the giving year.

4.2 Example of Analysis for a Group of Machines

The following six graphs (Fig. 9a–f) show functions for group of machines of average age 26 years. Figure 9a represents availability, Fig. 9b is use index related to age of machines. Figure 9c and d show cost $C(t)$ and profit function $P(t)$ also in age of machines, while Fig. 9e and f show the same function but related to its mileage.

Values for functions $C(t)$ and $P(t)$ have also been simulated for two and three shifts to verify profitability of organizational changes in operational management.

It is seen in Fig. 9a slight decrease in machine availability and a clear decline in the use index (Fig. 9b). Annual mileage of younger machines (10 years old) are at the level of 1700–2000 mth a year, then fall down to 1000 mth for oldest machines. The

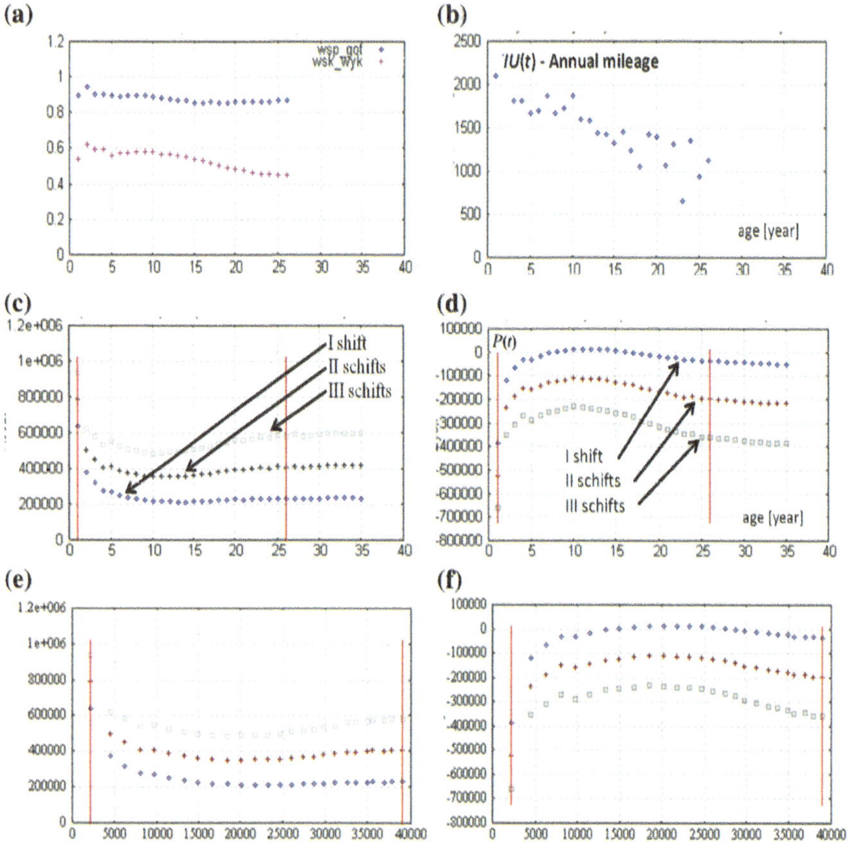

Fig. 9 Assessment functions for group of machines of average age 26 years. **a** Availability, **b** use index, **c** cost function $C(t)$ related to age, **d** profit function $P(t)$ related to age, **e** cost function $C(t)$ related to mileage, **f** profit function $P(t)$ related to mileage. (◊)—objective function value obtained assuming one-shift operation, (+)—objective function value obtained assuming two-shifts operation, (•)—objective function value obtained assuming time three-shifts operation, | |—vertical lines define the range of the age/mileage for suitable period for machine exchange (withdraw from operation)

study of both objective functions shows existence of minimum costs and maximum profits for the machines aged 10–15 years.

Due to the low utilization of equipment after 10 years (long repairs) and a higher purchase price, function $P(t)$ (Fig. 9d) only slightly exceeds the threshold of cost-effectiveness and is negative for machines older than 18 years. It is the reason why machines should be exchanged 15–20 years of operation, immediately as they leave a range of minimum cost and maximum profit (18,000–20,000 mth).

5 Summary

Paper presents problems and various approaches to assessment earth-moving machines. Important factors influencing that assessment cover operation and maintenance management, machine quality and cost, spare parts and materials availability.

Anyway, any assessment is only a support to undertake operational decision concerning purchase, withdraw from operation or choosing operational strategy. That is why assessment criteria have to be selected carefully according to assessment objectives.

It was shown as a simple assessment based on cost and reliability and complex one including costs, usage, reliability. Example of a set of loaders assessment has confirmed the applicability of the approach supporting decision of effective machine fleet management.

References

1. IEC 60050–191: Dependability and quality of service. Amendments 1 & 2. International Electro Technical Vocabulary (IEV 191). Based on IEV on-line. http://domino.iec.ch/iev
2. Karpiński, J., Firkowicz, S.: Zasady profilaktyki obiektów technicznych. PWN, Warszawa (1981)
3. Młyńczak, M., Wiewiórko, B.: Złożone systemy maszynowe w analizie niezawodności bezpieczeństwa. Konferencja Przeglądowa Programów Badawczych KBN, Radom (1996)
4. Barringer, H.P.: Life cycle cost and good practices. In: NPRA Maintenance Conference Proceedings. Barringer&Associates, Inc. San Antonio, Texas (1998)
5. Barringer, H.P., Weber, D.P.: Life cycle cost tutorial. In: Fifth International Conference on Process Plant Reliability. Gulf Publishing Company, Huston (1997)
6. Boyer, K.K., Pagell, M.: Measurement issues in empirical research: improving measures of operations strategy and advanced manufacturing technology. J. Oper. Manage. **18**, 361–374 (2000)
7. Dhillon, B.S.: Maintainability, Maintenance, and Reliability for Engineers. CRC Press. Taylor & Francis Group, Boca Raton (2006)
8. Młyńczak M.: Analiza kosztów eksploatacyjnych z uwzględnieniem niezawodności obiektu. Materiały XXVI Szkoły Zimowej Niezawodności. Sekcja Podstaw Eksploatacji, Polska Akademia Nauk, Szczyrk (2007)
9. Pesonen, L.T.T.: Implementation of design to profit in a complex and dynamic business context. University of Oulu, Finland, Oulu (2001)
10. Żółtowski, B., Niziński, S.: Zarządzanie eksploatacją obiektów technicznych za pomocą rachunku kosztów. MARKAR – BŻ, Zielonka (2002)
11. Hansen, D.R., Mowen, M.M., Guan, L.: Cost management. Accounting and control. South-Western Cengage Learning, Mason (2009)
12. Turney, P.B.B.: Activity-Based Costing an Emerging Foundation for Performance Management. Cost Technology, Inc. www.sas.com/resources/whitepaper/wp_5073.pdf. Accessed 11.2011
13. Wallen, A.: Activity-Based Costing. US Air Force. Web page: www.dau.mil/conferences/presentations/2002/briefings/S2T6-ABC_AndrewWallen.pdf. Dostęp: 11 2011
14. Olearczuk, E.: Zarys teorii użytkowania urządzeń technicznych. WNT, Warszawa (1972)
15. Ayyub, B.M., Klir, G.J.: Uncertainty Modeling and Analysis in Engineering and the Sciences. Chapman & Hall/CRC, Taylor & Francis Group, Boca Raton (2006)

16. Garvey, P.R.: Probability Methods for Cost Uncertainty Analysis: A Systems Engineering Perspective. Marcel Dekker Inc., New York, Basel (2000)
17. Webpage of Krakow Economic University: http://mfiles.ae.krakow.pl/pl/index.php/Amortyzacja Kraków. Accessed 12.2010
18. Russell, R., Taylor, B.W.: III, Operations Management, 5th edn. Wiley, Inc., University of Tennessee at Chattanooga (2006)
19. Slack, N., Chambers, S., Johnston, R.: Operations Management, 4th edn. Pearson Education, Harlow (2004)
20. Młyńczak, M.: Safety analysis in operation and design of open pit machines. advances in safety and reliability. In: Proceedings of the ESREL'97, International Conference on Safety and Reliability. Elsevier Science, Pergamon, Lisbona (1997)
21. Żółtowski, B., Niziński, S.: Modelowanie procesów eksploatacji maszyn. MARKAR-BŻ. Białe Błota (2002)
22. The Fast Guide to OEE™: Vorne Industries Inc., Itasca, IL USA. www.vorne.com, www.oee.com. Accessed 03 2012
23. Młyńczak, M.: Metodyka badań eksploatacyjnych obiektów mechanicznych. Monografia. Oficyna Wydawnicza Politechniki Wrocławskiej, Wrocław (2012)
24. Kletz, T.: HAZOP and HAZAN. Identifying and assessing process industry hazards. Institution of Chemical Engineers, Rugby (1992)
25. Gołąbek, A., Młyńczak, M., Nowakowski, T.: Podniesienie poziomu niezawodności i gotowości eksploatacyjnej ładowarek. Prace Naukowe Centralnego Programu Badań Podstawowych 02.05. Wyd. Politechniki Warszawskiej. Warszawa (1990)
26. Jodejko-Pietruczuk, A., Młyńczak, M., Zając, M.: Assessment of economical lifetime of heavy-duty machines. Case study. In: Proceedings of the European Conference on Safety and Reliability, ESREL 2009. A.A. Balkema, Monachium, Praga (2009)

Dynamical Loadings of Booms in Front-End Wheel Loaders

Wiesław Bereś and Kazimierz Pieczonka

Abstract The paper discusses the problem of dynamic loadings of the boom components in a front-end wheel loader during discharging of the bucket. A discrete two-dimensional model of a front-end loader on tyres was developed for numerical analyses of the loader boom dynamics. The mass of the entire machine including the mass of bulk material transported was reduced to three masses with five degrees of freedom (5-DOF). The masses are connected using the stiffness and damping elements. On the basis of this physical model, a mathematical model of the dynamic system was created. The final system of eight linear differential equations with time-dependent coefficients is solved using the Runge–Kutta method of the fourth-order. A computer program WYS1 that solves the mathematical model numerically was written using the ALGOL 1900 computer language. In addition, a simplified method is presented that can be used to calculate dynamic loadings of the boom system in front-end wheel loaders on tyres.

Keywords Earth-moving machinery · Front-end loaders · Booms · Dynamic analysis · Dynamic models

1 Introduction

One of the main challenges in design of the front-end wheel loaders is proper design and dimensioning of the boom that operates the loader bucket.

Kazimierz Pieczonka: Passed away on August 3, 2017.
This is an abridged translation of the original paper published in Polish in Przegląd Mechaniczny, 1977, Vol. XXXVI, No. 20, pp. 703–706. Information contained in this paper were current in 1975 but now may be considered obsolete.

W. Bereś (✉)
Faculty of Mechanical Engineering, Wroclaw University of Sciences and Technology, ul. Wybrzeze Wyspianskiego 27, 50-370 Wroclaw, Poland
e-mail: wieslaw.beres@nrc-cnrc.gc.ca

NRC-Aerospace Research Centre, Ottawa, ON K1OR6, Canada

M. Sokolski (ed.), *Mining Machines and Earth-Moving Equipment*,
https://doi.org/10.1007/978-3-030-25478-0_6

Recently, a lot of attention in design of earth-moving machinery has been directed towards assessment of the cutting forces in various types of excavators as well as assessment of loading resistances for loading of various bulk materials in different types of loaders. However, the problem of dynamic loadings of boom components during discharging of the bucket was not treated so far with sufficient attention. Typical bucket discharging process involves deceleration of substantial masses rotating with significant initial rotational speeds in a short period of time. Resulting deceleration causes significant impact loadings that may create component damage and consequently negatively affect reliability and operability of the loader. In particular, repeated impact loadings may cause quick progressing damage of the boom components, mainly bushings in boom component connectors. In addition, these repeated impact loadings may cause plastic deformation of the rod that opens and closes the bucket thus leading to a major failure of the entire boom system due to excessive backlash in the system and breaking down the boom kinematics.

Therefore, the paper describes a method of establishing dynamic loads acting on a front-end loader boom during discharging of the bucket. To establish the basic characteristic of the discharging process, a high-speed camera was used and discharging was filmed for a front-end loader transporting gravel. Using this preliminary exploratory research, two types of bucket discharging processes were distinguished. In the first discharge type, the boom with the full bucket is raised by two hydraulic cylinders and then the hydraulic system controlling the bucket is activated through a swing arm. As a result, the bucket rotates with respect to the boom and the bulk material slowly leaves the bucket under gravity. At the same time, due to change of the location of the centre of gravity of the entire loader caused by decrease of the total mass, the vibrations of the boom as well as the entire loader are excited. It could be clearly observed in the movie recorded during this preliminary research that the entire loader vibrated on the tyres, which are clearly the elements with the lowest stiffness in the whole dynamical system. In the second type of discharge, which happens when the loader boom is in its highest position, two separate periods may be distinguished. The first period is similar to the discharge process described previously but in this case the bucket is not fully emptied. This period ends at the time point in which the swing arm contacts the stoppers which prevent the boom members from reaching critical, "dead" positions. The second period starts at this point of time. In this period the kinetic energy of the bucket together with the mass of bulk material transported in the loader is converted into potential energy of deformation of the boom elements, the tyres, as well as into potential energy of the fluid in the boom hydraulic system.

2 Physical Idealization of a Front-End Wheel Loader

A discrete two-dimensional model of a front-end loader on tyres was chosen for numerical analyses of the loader boom dynamics. It was assumed that:

- the loader is symmetrical with respect to a vertical plane that passes through the geometrical centre of the machine;
- the loader is in the position to drive straight ahead; and
- the loader is located on a horizontal surface.

These assumptions were made on the basis of the previous works on dynamic analyses of other types of vehicles such as research on ride dynamics of automotive vehicles.

A scheme of a multi-body model of a front-end loader on tyres is shown in Fig. 1. The mass of the entire machine including the mass of bulk material transported was reduced to three masses with five degrees of freedom (5-DOF).

These three masses are:

- Bucket with bulk material; the total mass m_3, the centre of gravity S_3 and the central moment of inertia I_3.
- Boom with the total mass m_2, the centre of gravity S_2 and the central moment of inertia I_2, and
- Reminder part of the machine with the total mass m_1, the centre of gravity S_1, and the central moment of inertia I_1.

The spring elements representing the stiffness of tyres in the direction perpendicular to the surface are denoted c_1, and c_2; subsequently c_3 represents the stiffness of all tyres in the direction parallel to the surface; c_A is the torsional stiffness of the boom with respect to the body of the machine, and c_M is the torsional stiffness of the bucket with respect to the boom. The recent stiffness is acting only when the bucket start discharging the bulk material and when the bucket is in the contact with the

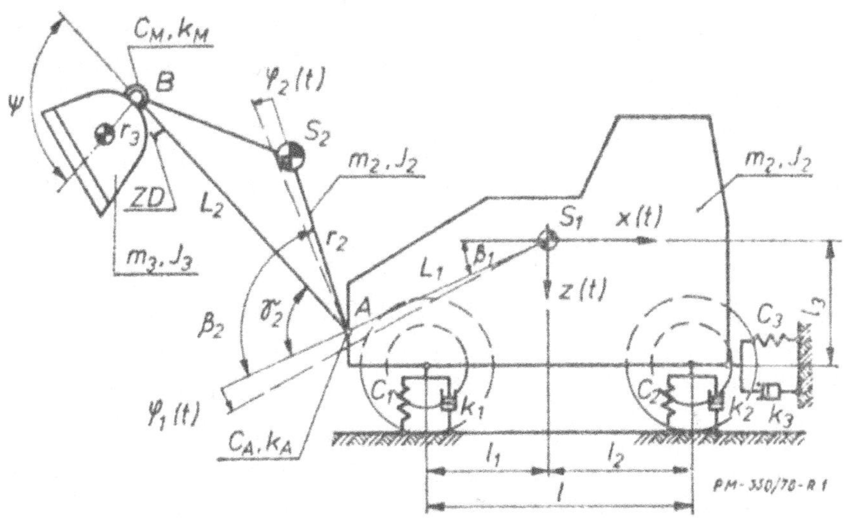

Fig. 1 Dynamic model of a front-end wheel loader

stopper installed at the arm of the boom. The damping elements k_1, $k_{2,}$ and k_3 are assumed to act in parallel to the stiffness elements as shown in Fig. 1.

3 5-DOF Mathematical Model of the Wheel Loader

On the basis of the physical model described, a mathematical model of the dynamic system was created. A vector of the generalized coordinates has a shape:

$$q_1(t) = [x(t)z(t)\phi_1(t)\phi_2(t)]^{\mathrm{T}} \quad (0 \le t \le t_w) \tag{1}$$

$$q_2(t) = [x(t)z(t)\phi_1(t)\phi_2(t)\phi_3(t)]^{\mathrm{T}} \quad (0 \le t \le t_w) \tag{2}$$

where

- x and z are horizontal and vertical displacements of point S_1 with respect to the ground, which is considered to be the reference system,
- φ_1 is the torsional displacement of mass m_1 with respect to the initial position
- φ_2 is the torsional displacement of mass m_2 with respect to mass m_1 in point A,
- φ_3 is the torsional displacement of mass m_3 with respect to mass m_2 in point B, and
- T denotes the transpose operation of the matrix.

Equation (1) relates to the first period of the discharge, while Eq. (2) describes the second period of the discharge process described in the Introduction.

In the first period of the bucket discharge process, the enforcing function—generalized displacement—is the change of the angle ψ describing the position of the bucket with respect to the boom. This function depends on time, $\psi(t)$. It was assumed that the function $\psi(t)$ is continuous and has continuous derivatives of the first and the second order. To take into account a general case, it was assumed that the dynamic parameters of the bucket with bulk material can change in time, i.e.,

$$m_3 = m_3(t), \quad I_3 = I_3(t), \quad r_3 = r_3(t) \quad (0 \le t \le t_w) \tag{3}$$

where t_w is the time of the first period of the discharge process, that is the time from the process beginning until the swing arm comes into contact with the stopper. If the displacement of the hydraulic piston is s_w, and its speed is constant v_0, t_w is obtained as

$$t_w = \frac{s_w}{v_0} \tag{4}$$

To obtain equations describing motion of the machine in the first and the second period of discharge, the kinetic energy, potential energy, and dissipation function were expressed as functions of the generalized coordinates and generalized velocities of the system. It was assumed that vibration of the machine with respect to the equilibrium position is small, therefore the expression for kinetic energy was linearized, and then the Lagrange equations of the second type were used [1]. It allowed for derivation of the equations of motion which in matrix notation have the shapes as

- for the first period

$$A_1(t) \cdot \ddot{q}_1(t) + B_1(t) \cdot \dot{q}_1(t) + G_1(t) \cdot q_1(t) + p(t) = 0 \tag{5}$$

- for the second period

$$A_2(t) \cdot \ddot{q}_2(t) + G_2(t) \cdot q_2(t) + p(t) = 0 \tag{6}$$

The elements of matrices A_1, B_1, G_1 and p depend on the parameters of the dynamical system as well as on-time t. However, the element of matrices A_2 and G_2 depend only on the parameters of the dynamical system.

It was assumed that at the beginning of the first period of the discharge the machine is not moving and all its components are located at the beginning of the coordinate systems. That means that the initial conditions have the shape:

$$q_1(t) = 0, \quad \dot{q}_1(t) = 0 \tag{7}$$

while for the second period the initial conditions were calculated from the state of the dynamic system at time t_w for the first period:

$$q_2(0) = [q_1(t_w), 0]^{\mathrm{T}}, \ \dot{q}_2(0) = [\dot{q}_1(t_w), \omega_0]^{\mathrm{T}} \tag{8}$$

In the second period of the discharge, the time shift was introduced as $t \to t - t_w$. In (8) ω_0 is the instantaneous angular velocity of the bucket with respect to the boom at the moment in which the swing arm is coming into contact with the stopper. The angular velocity ω_0 was established using the method and the software described in [1].

To solve the systems of differential equations (5) and (6), they were converted by introducing a new vector of unknown that included both the generalized coordinates and generalized velocities

$$y(t) = [\dot{q}_1(t)\ q_1(t)]^{\mathrm{T}} \tag{9}$$

The final system of eight linear differential equations with time-dependent coefficients was solved using the Runge–Kutta method of the fourth-order [2, 3].

Using the algorithm described, a computer program WYS1 was written using the ALGOL 1900 computer language. For simplification of calculations, a worst-case scenario was assumed and the time-dependent parameters m_3, I_3, and r_3 were assumed to be constant. The function ψ(t) and its time derivatives were given as the polynomials described in [1]. It was assumed that $v_0 = \text{const} \neq 0$. The numerical integration procedure for the system of differential equation used is described in [4].

Program WYS1 uses the parameters of a wheel loader model (m_1, m_2, I_1, I_2, R_2, β_1, β_2, γ_2, L_1, L_2, k_1, k_2, l_1, l_2, l_3, c_1, c_2, c_3, s_w), coefficients of the polynomials describing the time changes of ψ(t): a_i ($i = 0, 1, \ldots, k$) and the velocity of the hydraulic piston v_0 and integrates the system of differential equation (5) with zero initial conditions in the time period of (0, t_w). The results are obtained in a form of a table containing values of the generalized coordinates x, z, φ_1, φ_2 and their derivatives at twenty discrete points of time. The algorithm—block scheme—of the program WYS1 is shown in Fig. 2.

Subsequently, a solution of the system of Eq. (6) is sought in a shape of [5, 6]:

$$q_2(t) = \psi \cdot \sin(\omega \cdot t + \vartheta) \tag{10}$$

and therefore

$$(G_2 - \omega^2 \cdot A_2)\psi = 0 \tag{11}$$

after assuming $G_2 = L \cdot L^{\mathrm{T}}, x = L^{\mathrm{T}} \cdot \psi$, where L is a triangular matrix, the problem is reduced to searching of the eigenvalues and eigenvectors of the matrix

$$\boldsymbol{S} = \boldsymbol{L}^{-1} \cdot \boldsymbol{A}_2(\boldsymbol{L}^T)^{-1} \tag{12}$$

Finally, a solution to the matrix Eq. (6) has a shape

$$q_2(t) = \sum_{i=1}^{n} D_i \cdot \psi^{(i)} \cdot \sin(\omega_i t + \vartheta_i), \;\; (n = 5) \tag{13}$$

The initial conditions expressed in (8) allow for calculation of $2n = 10$ constants $D_1, D_2, \ldots, D_n, v_1, v_2, \ldots, v_n$. Finally, the time trajectory of the dynamic load on the boom looks like:

$$q_3(t) = \sum_{i=1}^{n} D_i \cdot \psi_n^{(i)} \cdot \sin(\omega_i t + \vartheta_i), \;\; (n = 5) \tag{14}$$

A question of finding the maximum dynamical loading of the boom is tantamount to finding the first local maximum of the function $\varphi_3(t)$. Calculating derivatives of Eq. (14) and finding numerically the time $t_{\max}$ at which the first maximum of the function occurs, the maximum dynamic loading of the boom is:

$$M_{max} = C_M \cdot \phi_{3max} \tag{15}$$

Using the algorithm presented here the computer program WYS2 was created using AGOL 1900 computer language. The program WYS2 reads the initial

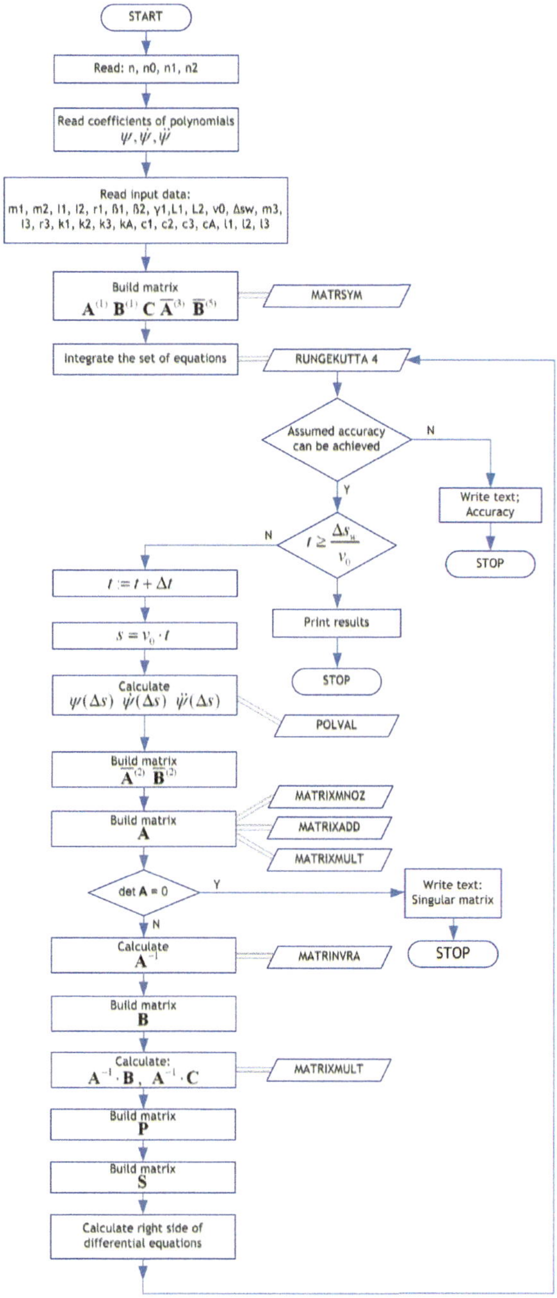

Fig. 2 Algorithm of the program WYS1

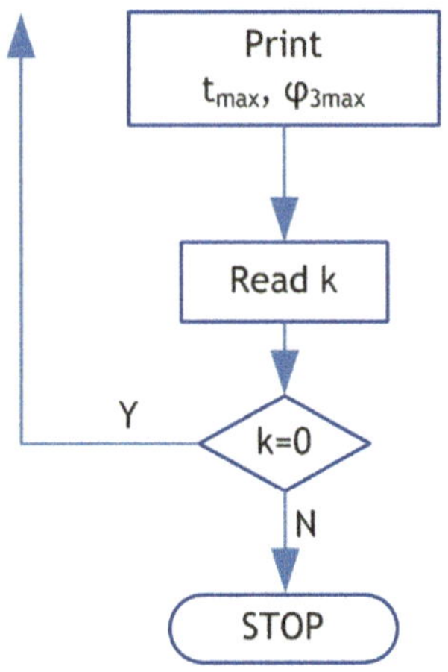

Fig. 3 Block diagram of the program WYS2

conditions and calculates both values tmax and $\varphi_{3\max}$. To calculate eigenvalues and eigenvectors of the symmetric matrix *S*, the algorithm CPEIGENSP described in [4] was used. The basic block diagram of the program WYS2 is shown in Fig. 3.

4 One-Degree-of-Freedom (1-DOF) Model

To simplify the calculations and to check the effect of the various parameters on the dynamic loading of the boom, the simplified, one-degree-of-freedom model of the bucket with bulk material was also prepared, as shown in Fig. 4.

In this model, the mass of the bucket with the bulk material was reduced to the mass *m* with the central moment of inertia *I*. The bucket is connected to the rigid boom through the torsional element with the torsional stiffness c_M. A generalized

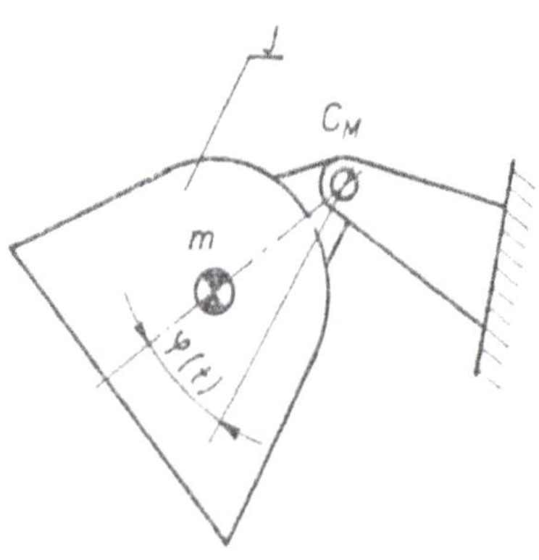

Fig. 4 1-DOF model of the system

coordinate φ describes the angle of rotation of the bucket with respect to the boom; the value of this angle is directly related to the dynamic loading of the system.

The mathematic model of this system has a shape of a linear differential equation

$$I \cdot \ddot{\phi}(t) + c_{\mathrm{M}} \cdot \phi(t) = 0 \tag{16}$$

with the initial conditions

$$\phi(0) = 0, \quad \dot{\phi}(0) = \omega_0 \tag{17}$$

A solution of this equation is

$$\phi(t) = \frac{\omega_0}{\omega} \sin(\omega \cdot t), \quad \omega = \sqrt{\frac{c_{\mathrm{M}}}{I}} \tag{18}$$

The maximum value of the angle φ corresponding to the maximum loading of the boom system in the loader is obtained at a time point time of

$$t_{max} = \frac{\pi}{2}\sqrt{\frac{I}{C_M}} \tag{19}$$

and this maximum has a value of

$$\phi_{max} = \omega_0 \cdot \sqrt{\frac{I}{C_M}} \tag{20}$$

Now, the maximum load for the boom system can be found using Eq. (15).

5 Approximation of 5-DOF Model with 1-DOF Model

The algorithms presented and program WYS1 were used to perform numerical calculations and obtain time trajectories of the generalized coordinates x, z, φ_1, φ_2. The values of parameters corresponding to a scale model of the front-end wheel loader built at the Institute for Machine Design and Machine Operation at the Wroclaw University of Technology were used as the input parameters to analyses. The generalized loading function ψ(t) was specified through entering the coefficients of the approximating polynomials [1]. The calculations were performed for two fillings of the bucket, full and empty, and using two velocities of the hydraulic cylinder in the hydraulic system of opening and closing the bucket.

Figure 5 shows typical time changes of the coordinate $x(t)$, which represents the horizontal displacement of the centre of gravity for the entire loader. Four curves in

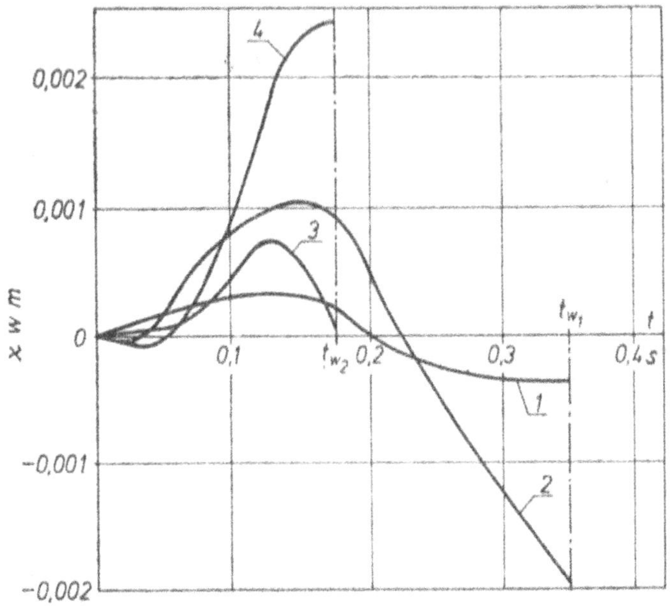

Fig. 5 Component $x(t)$—horizontal displacement with time in the first period of discharge process

this figure represent (1) empty bucket, (2) full bucket, piston velocity 0.19 m/s; (3) empty bucket, (4) full bucket, piston velocity 0.38 m/s. All responses were aperiodic.

For all trajectories obtained, increase the mass of the bucket resulted in an increase of dynamic loading on the tyres and on the hydraulic system. It has been also found out that increasing the velocity of the piston v_0 by a factor of two resulted in larger than the factor of two increases in the dynamic loading of the tyres and the hydraulic system.

From the time trajectories calculated the values of the generalized coordinates x, z, φ_1, φ_2, and their derivatives were obtained for time t_w. These values were used as the input for program WYS2. This second computer program analyzed the time domain trajectory of the generalized coordinate φ_3(t) which is directly related to the dynamical loading of the wheel loader boom system.

These calculations were repeated for several values of the wheel loader parameters. The results obtained for the 5-DOF system were compared to those obtained for 1-DOF system. The comparison showed that the responses of the major generalized coordinate φ_{3max} differed by 7–18%. It was also found that the responses for 1-DOF system were always larger than those obtained for 5-DOF systems.

Figure 6 shows a comparison of the time trajectories for both of these models. The coefficient of the quality of the approximation χ was defined as:

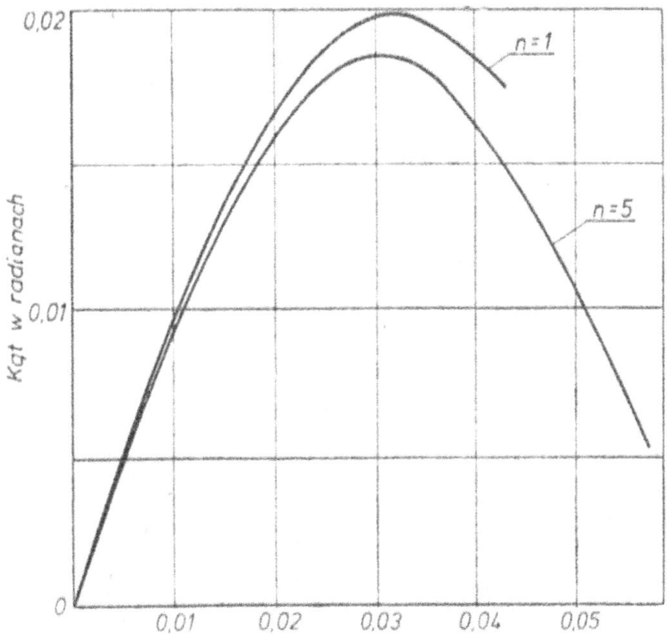

Fig. 6 Comparison of the response $\varphi_3(t)$ for two models: 1-DOF and 5-DOF. Initial rotational velocity of the bucket $\omega_0 = 1$ m/s, $c_M = 1230$ Nm/rad; $I = 0.5$ kg m^2

$$\chi = \left| \frac{\phi}{3\max \phi^{(1)}_{3\max} {}_{(1)^{(5)}_{3\max}} \;||} \right| \tag{21}$$

It was concluded that the much simpler 1-DOF model can approximate a more complicated 5-DOF model with the coefficient of the quality of the approximation ranging between 15 and 20%.

6 Dynamic Loading of Wheel Loader Booms During the Bucket Discharge Process

On the basis of the analyses performed, it is suggested to calculate the largest dynamic loading of boom systems for front-end wheel loaders on tyres as follows:

$$\phi_0 \sqrt{\frac{I}{c_{M_{3\max}}}} \tag{22}$$

where ω_0 is the instantaneous angular velocity of the bucket when the bucket stopper engages in contact with the swing arm, in rad/s; c_M is the torsional stiffness of the bucket with respect to the boom in Nm/rad, I is the moment of inertia of the boom with bulk material transported with respect to the boom axis of rotation in kg m^2. Using Eq. (15) the maximal dynamic forces acting on the boom can be calculated.

After some modifications, the methods presented can be also used to establish dynamic loadings of the boom systems of other earth-moving machinery, such as excavators or front-end wheel loaders for mining application.

Note

During the preparation of this paper, I learned, with great sadness, about the passing of my Ph.D. thesis supervisor, my mentor, my co-author and later a personal friend, Prof. Kazimierz Pieczonka. "Kaziu" as we colloquially called him at the Wroclaw University of Technology, was a great researcher and a renowned specialist on the topic of earth-moving machinery. In addition to being a very decent human being, Kaziu was always calm and smiling, very open to people and never held any grudges. I will always remember his intellect, vision, and leadership. His passion and enthusiasm for work, focus on his endeavors combined with common sense when approaching life's challenges were passed on to many generations of young and not-so-young researchers. Professor Pieczonka will be remembered as an outstanding scientist, a faithful friend, and a respected colleague. He will be missed by many. WB.

References

1. Bereś, W., Pieczonka, K.: Wyznaczanie parametrow kinematycznych wysiegnikow ladowarek lyzkowych (A method of establishing of kinematic characteristics of the bucket loader booms) (in Polish), Vol. XXXVI, No. 13, pp. 455–458. Przeglad Mechaniczny (1977)
2. Demidowicz, B.D., Maron, I.A., Szuwalowa, B.J.: Metody numeryczne, czesc II (Numerical methods, part II) (in Polish). PWN, Warszawa (1965).
3. Ralston, A.: Wstep do analizy numerycznej (A first course in numerical analysis) (in Polish). PWN, Warszawa (1965)
4. ODRA-ALGOL: Procedures for the computer ODRA 1204, part II, 1204-VIII-11. WZE Elwro, Wroclaw (1970)
5. Bishop. R.E.D., Gladwell, G.M.L., Michaelson, S.: Macierzowa analiza drgan (The matrix analysis of vibration) (in Polish). WNT, Warszawa (1972)
6. Gutowski, R.: Mechanika Analityczna (Analytical Mechanics) (in Polish). PWN, Warszawa (1971)

Modern Design of the Transport Vehicles Drive Structures

Franciszek W. Przystupa

Abstract In the modern design of the transport vehicles drives—universal joints are frequently derived from the synchronous operation. In the paper were discussed ways to counteract this phenomenon. It is necessary to build a diagnozer for the evaluation of the changes in the rigidity of the supports, internal dumping, rigidity of components, etc. Method of diagnosis will allow an assessment of the spatial position of the shafts, their dynamics and as a result, adjust the phase of the shaft. Phase control in real-time will occur through the use of smart tools.

Keywords Vehicles · Drive · Universal joints · Spatial position of the shafts

1 The Technical Situation

The main trends in the vehicles design are directed to optimization of:

- Engine power,
- Engine torque and engine speed—optimization of gears,
- Axle transmission ratios,
- Weights,
- Sound insulation and acoustic phenomena,
- Environmental protection.

In order to achieve these objectives in the drive system of transport machinery and any other self-propelled vehicles, it is necessary to synchronic transfer of power (torque and speed) from the engine to the driven wheels and other components. It is necessary to apply shafts with angular CV (constant velocity) joints, conveying motion and torque. The constant velocity joints, synchronous or synchronizing commonly referred to as a universal joint, regardless of the angle of shaft parts should rotate at the same angular speed without the occurrence of rotary motion and the

F. W. Przystupa (✉)
Faculty of Mechanical Engineering, Wroclaw University of Science and Technology, ul. Wybrzeze Wyspianskiego 27, 50-370 Wroclaw, Poland
e-mail: franciszek.przystupa@pwr.edu.pl

M. Sokolski (ed.), *Mining Machines and Earth-Moving Equipment*,
https://doi.org/10.1007/978-3-030-25478-0_7

torque of the drive serious pulsation. Lack of this condition will cause the drive machine vibration, noise and accelerated degradation [1–25].

In the machinery and in the vehicles, it is necessary to power transmit without distortion from the engine to the wheels and other powered components. For this purpose, the shafts with universal joints, CV or synchronous joints—known as Cardan joints—are necessary. Such joints regardless of the angle of the between of shaft pars—driving and driven—rotates at the same angular speed, without the presence of dangerous rotation pulse and torque changes.

The problem concerns in transporting machines with different pre-dynamic characteristics (between successive copies of the machines) or the changing characteristics according to operation—for a specific machine.

Shafts and axles and drive shafts of the joints, generally referred to as synchronous, universal or Cardan joints must be smooth, small, lightweight and meet all other criteria of the final manufacturer—the vehicle manufacturer. The shafts with joints should not cause additional vibration in the drive—requires even composition elements that effectively dampen vibrations. Producers of shafts, axles and axle shafts solutions should be able to implement a number of auxiliary tasks, beyond a basic task—drive synchronous transmission, e.g. the oscillation frequency of the drive damping or control by adapting to the specific vehicle vibration [1–34].

Among the principal objectives can be mentioned minimization or optimization:

- Joints rotational phase (instantaneous angle of rotation—Phase Adjusting),
- The full effect of NVH (noise—vibration—harshness),
- The dynamic effects of other parts,
- The balance,
- The weight reduction,
- The state observation,
- The effectors, etc.

These different objectives can be achieved through various methods: the movement and limiting masses, kinematic regulations, the use of smart tools, etc.

For example—in [28], it stated that it is possible to design the shaft of the hybrid composite fibre pipe and aluminum pipe with a smaller mass in comparison with the steel shaft the shaft with the primary natural vibration frequencies above the shaft speed. The paper [29] shows that the use of composite material, Fig. 1, greatly reduces the weight of the shaft. The initial torque required to obtain the drive speed is high, weight is reduced; the excess torque is used to propel the vehicle and not

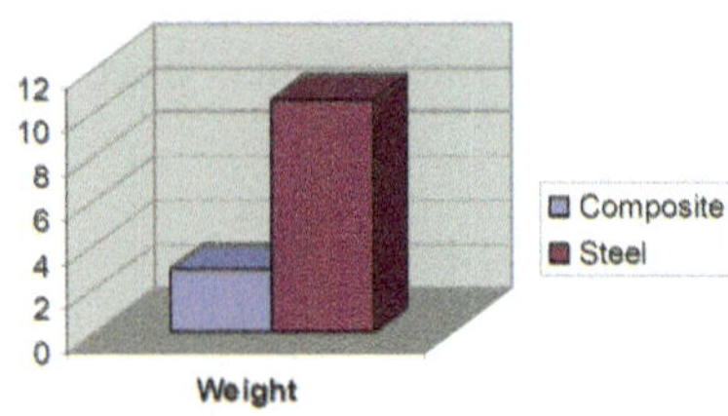

Fig. 1 Mass comparison between steel and composite shaft (kg)

to overcome the inertia of the rotating shaft portion. Reducing the weight of shaft increases the frequency of the deflection and therefore the shaft can be used at higher frequencies than the steel shaft. The reduction in weight gives additional benefits in the form of increased fuel efficiency.

Other aspects discussed work [30], where it is analyzed according to production costs and angles of Cardan-type joint action. The results indicate that the increase in the angle of refraction in the joint requires a corresponding increase in production costs. However, a significant reduction in production costs could be realized by reducing the angle to a value below 30°.

It is used ceramics smart [11, 31], introduced a solution containing the structure of the shaft of carbon fibre (CARBON FIBER DRIVE Shaft)—Fig. 1 extra vibration dampers (DRIVE SHAFTS Damper) or retarders, referred to as self-regulators vibrations using magnetorheological effect, and so on.

The structures have built-in embankments with radical changes, acting on behalf of the dynamics of vehicle movement—e.g. torsional vibration dampers or retarders, assembled in different configurations, using different methods of dispersing or recuperation. They shall also resolve incredibly subtle—e.g. modified in several ways motor spline connection, modified bearing joints, shaping to pass geometry shafts, splines of linear bearings, etc.

The ability to carry out particular tasks and goals can be achieved through a number of essential partial tasks. One method uses different regulations gelding through "smart" tools. The next will present some aspects of the use of this tool, and in particular to obtain the vehicle synchronicity drive.

The indicated tasks and targets can be achieved through many intermediate steps. One of the methods uses varied smart tools. Some aspects of this tool, in particular for synchronicity in machine power transmission will be shown.

2 Synchronous Drive Transmission

In order to achieve drive machine synchronicity—compliance angular velocity of the input and output shaft, or a constant angular velocity of the input shaft of an axle (shaft input), universal joints of the drive shaft must meet known requirements: two joints angle of refraction α of one shaft should be the same. All parts of the shaft (before the joint between the joints for the articulation shafts 1, 2, 3) should lie in the same plane (vertical plane parallel to the longitudinal axis of symmetry of the machine cradle of two joints of one of the shaft (O and A) must lie in one plane. Condition extra: an input shaft (1) and output (2) should be parallel to each other—usually carried out in a classic drive system.

In fact, despite attempts to ensure the fulfilment of the synchronicity conditions methods of construction in the structure of the machine, in the machine structures, there are the phenomena of deformation (α angles, and movement), leading shaft out from the state proper operation. Supporting shaft structure is founded spatially, with

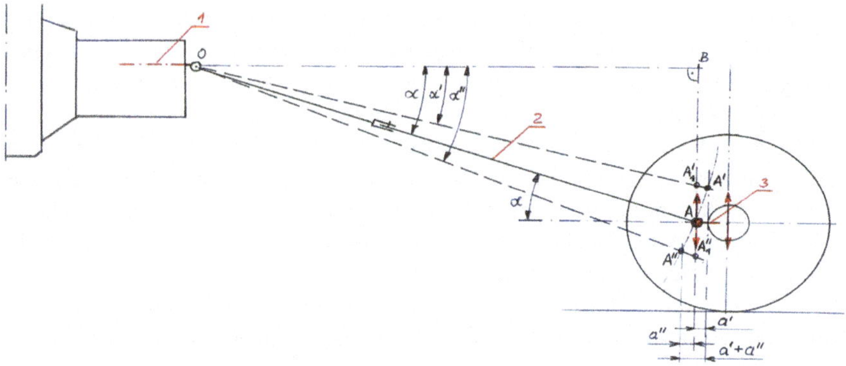

Fig. 2 The structure of the drive system in the vehicle

unpredictable parameters of the foundation, depending on the momentary technical situation. The drive is working out of synchronicity (Fig. 2).

When the drive shaft is set in the correct phase—that has preserved properly angle parameters—it is not a source of torsional vibrations. When the shaft is not in the correct phase—becomes a VBA generator—and a source of torsional vibrations.

In the accidental—spatial position of all the shafts it is possible to maintain synchronous power transmission by dynamically adjusting the phase shift of the joint of the Cardan shaft to the instantaneous position of the input and output shafts relative spatial position. As the phase shift is meant an angular displacement of the fork shaft intermediate the position of the single-leaf. This position is appropriate only single-leaf for the ideal state—meeting all three conditions of synchronicity.

Dynamic adjustment of the phase shift of the joint of the Cardan shaft to the instantaneous relative spatial position of the input and output shafts can be realized by adjusting of the phase between the joints due to the following requirements:

1. Synchronous input rotations from baseline to avoid kinetostatic fluctuations for determining the quality of the process and to avoid changes in the external load bearing shaft.
2. Synchronous input rotations from baseline to avoid dynamic effects in the drive and the driven machine, in particular, to avoid any resonances and clearance elimination by excessively high vibration amplitudes, and additionally.
3. Additional desynchronized rotation input and output to produce the dynamic effects in the drive and the driven machine, for damping vibrations generated outside the shaft in the drive system.

So far, this task was carried out by several methods—material or through the use of passive and active dampers. The last two methods rely on the liquidation of the negative effects of dynamic, leaving the cause unchanged. These methods counteract the effects without eliminating the basic phenomenon—asynchronous transfer of powers through the drive shaft. However, prevention of excessive hazards and the resulting damage and failure is much less expensive than the removal of their consequences. Vibration dampening is to adjust the frequency dumpers to the dynamics of the machine—requires diagnosing the power train, together with the type of Cardan shaft.

Methods involving the elimination of the causes of negative dynamic phenomena, generally require phase adjustment between the joints of the shaft, which must be preceded by specifying the type of the shafts elements spatial position and the growing dynamic effect.

3 Diagnostics Observation of the Shafts Elements Spatial Position and the Growing Dynamic Effect

In the follow-up diagnostic processes and positions necessary for the active phase adjustment shaft diagnosed with mileage, algorithms, dynamic, full and partial results positions of the shaft and the dynamic effects of temporary state planning. The aim is usually to control the prediction of the developing process to avoid undesirable conditions or operations optimal limits. It is necessary to build the driver state—including diagnoser—the PTO shaft. Diagnostic method should allow the assessment of the spatial position of the shaft, the shaft dynamics and ultimately phase adjustment shaft. Use of CAE methods to be here (E).

The selected method of diagnosis at the design stage of drive system in machinery to avoid the accelerated degradation—when there is no possibility of a preliminary verification of the unit operating machinery.

The necessity of this diagnosis results from the use of the system equipment and transport propshaft, in which the environment during the operation of change of stiffness supports, in shafts—changes suppression of internal stiffness of components by weight—during operation while the same shafts also have different characteristics, initial—for specific copies of the shaft.

Generally intervention using the method consists of the sequence of shifts from unacceptable state → action → acceptability, to the acceptable state. It is necessary to create algorithm:

1. Measurement of the spatial position and/or vibration portion of the system drive containing shafts and inertia
2. The parameters analysis of the spatial position and/or vibration in terms of frequency and amplitude (speed, acceleration)
3. Decide for the intervention phase of the shaft
4. Intervention in the characteristics of the phase
5. Monitoring of the effect of interventions undertaken and executed
6. Further adjustment phase, according to. previous algorithm (1–5) or abandonment of further intervention.

Further it is desirable to:

- Periodic planned to take measurement of the spatial position and/or vibration in accordance with the specified algorithm,
- or the introduction of monitoring the spatial position and/or vibration with built-in alarm system after exceeding the limit specified parameter vibration.

The measurement of the spatial position and/or vibration of the shaft system drive containing shafts and inertia should be implemented:

- In the production phase, and preferably during the transport machine application—is a possible intervention in the outer parameters of the drive system.
- Periodically during operation of—intervention in the internal phase parameters or external parameters of the power train.
- After exceeding the limit of specified parameter vibration (with monitoring system) with the existence of vibration—intervention in the internal phase parameters (the permissible level exceeding was to notice early enough that there was no need to intervene in the outer parameters of the drive system.

The introduction of the diagnostic system to the machine should allow for a positive change in the existing system, the current state of the drive system. Algorithm (simplified) requires:

- preparing of diagnoser (depending on the nature and level of positions and parameters to be observed or vibration)
- establish performance levels in optimal points of the machine
- recording of the measurements at various points
- analysis and inference
- regulatory decision—in the muffler or
- intervention decision—the drive system.

The differential diagnostic method being proposed by the Author (in the 1990th) permits in CV shafts to avoid labor-consuming classical methods, thus it permits to have important savings. The creation of a model based upon an object or process knowledge is avoided. It is sufficient to utilize known regularities concerning the creation of state boundaries.

In the proposed and technically verified method the possibility of a mutual diagnosing of objects and process is utilized, without the need of an a priori knowledge of the characteristics of their degradation. The state of the system is estimated only through various kinds of comparative coefficients, further named differential coefficients.

In classical methods of diagnosis, it is often difficult to separate symptoms of long-term overload from analogical symptoms resulting from a change of the state.

Both methods permit to compare the presented hereinafter block diagrams of models of the classical method (according to Cempel) and the differential method. In the differential method, there exists no model block, and the inference-decision block can also be simplified.

The application of the presented differential method in technical systems is based upon a periodical, comparative observation of characteristics that, according to assumptions, carry information about the state of the observation object (Fig. 3).

The Author's method concerning the diagnosis of the exploitation quality of the joints of the Cardan shaft of loaders can serve as a very generalized example. A single joint has four rotary bearings; in the shaft there are eight identical bearings, in the machine, there are 20 bearings. It is difficult to imagine a simultaneous failure of all the bearings, so it is a good object for such a type of diagnose (Fig. 4).

The thermic differential diagnosis (of the shaft with Cardan joints) consists in measuring the temperatures of selected bearings by means of a conventional method. The obtained values of temperatures T_i make it possible to compare the temperatures of the bearings of the joints in various differential configurations (observations of differences of temperatures $T_i - T_k$, of their quotients T_i/T_k, etc., e.g. $(T_1 * T_2) - (T_3 * T_4)$ (Fig. 5).

An arbitrary bearing with measured parameters that do not deviate from other, measurable and being measured parameters play the role of a reference standard. The diagnosis is resistant to instantaneous overloads, energetic shocks, etc., influencing the whole diagnostically observed system. An instantaneous overload is acting to the same degree on the elements being observed, the observer is not troubled by such an event.

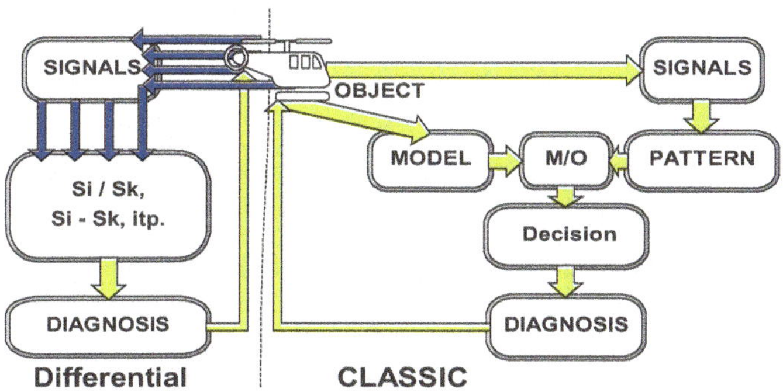

Fig. 3 Comparison of methods of diagnosis

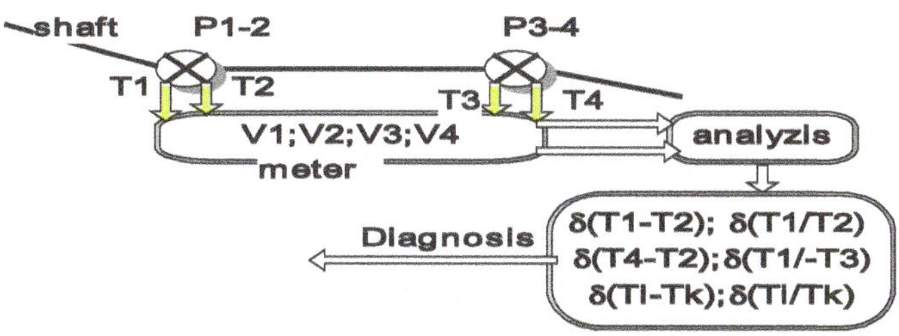

Fig. 4 Thermic differential diagnosis of the driving assembly (here a shaft with Cardan joints), T_i—temperatures, V_i—tensions

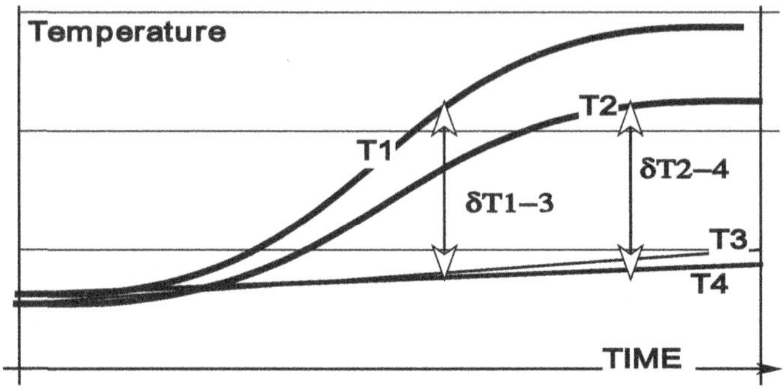

Fig. 5 Measurement of temperatures (T_i)—symptom of shaft failure. Measurements of T_1, T_2 are connected with a damaged bearing, of T_3, T_4 with a not damaged bearing—reference standard

For the case of necessity of overload observation, conventional methods are used. In order to apply the method in the case of an unknown object, it is necessary to:

- initiate the diagnosis on an efficient object, assure that, in the objects being observed, analogical conditions of exploitation technology exist,
- take into account the necessity of existence of analogical objects that function in series, in parallel or in analogical conditions,
- fulfil the conditions of a failure of a unique object, assure that the effort consequences (thermic) of failure are the same in the objects being observed,
- assure that the symptoms being observed carry good diagnostic information,
- assure that the symptoms being observed are really observable.

The introduction of the method into a real technical system requires the fulfilment of the mentioned conditions, and the execution of the following operations:

- the identification of objects for conditions of the differential method,
- exploitation measures of observability of symptoms of failure objects and not damaged objects,
- the determination of differential boundary values,
- the determination of periodicity of measurements,
- the choice of technical means—synthesis of diagnoser elements,
- the synthesis of the algorithm and analytical means of the diagnoser.

The method should be verified in particular machinery condition.

For the drive line with CV shafts—the following assemblies exist:

- analogical, identical or identical from the multi-criterion point of view, among them from the construction point of view,
- driven by the same power stream,
- driven by different power streams (parallel, mixed)

There is a potential of creation:

- of reference standards, their interchanging and replacing, rejecting
- of redundancy,

for the following situations:

- uncertain
- confirmation of inference logics
- overloads, changes in trends and of their creation
- characteristics of degradation, failures, and catastrophes.

One can propose the use of tools especially for image comparison, learning—with incorporation of dedicated media as magnets, gravimeters, scales, delamination meters, cameras, microphones.

Some typical reasons for VBA problem can be listed. Rear-wheel-drive driveshafts spin very fast and can vibrate. This will be felt in the seat, rather than the steering wheel. Acceleration and deceleration will often affect such a vibration. Driveshaft vibration results if there are tight universal joints, any dent in the tube or an improper shaft angle.

Watching the drive-shaft rotate will often reveal the problem. Any noticeable run out will cause a vibration. Removing the shaft and moving the universal joints is a good way to check them. Any resistance to motion, in either plane or slack in the joint means it is bad.

In another Detail Topic, we discuss drive-shaft angles. A simple test is to place a shim, under the transmission mount or the centre carrier bearing. This changes the angle of the shaft. A drive-shaft problem exists if the vibration changes. Checking the balance of a drive-shaft is best left to a drive-line specialist with the necessary equipment. Front-wheel-drive axles turn very slow by comparison. These will not generally cause a noticeable vibration unless they are badly bent. An exception is a bad inner C.V. joint, which will cause a vibration on acceleration and will quit on deceleration.

4 Tools for Actual Active Phase Adjustment in Real-Time

There are several tools for the smart correction of active (reactive) phase adjustment, in real-time, allowing the reaction to potential adverse dynamic. Mentioned earlier regulation or optimization are necessary for improving phase of the rotary joints (Phase Adjusting), the full effect of NVH, the dynamic effects of other (Dynamic Effects), balance (Balancing), the change in mass (Weight Reduction), and observations (State Observation) or the position of effectors (Effectors), etc.

Smart characteristics elements may be introduced as elements integral with the shaft, or as cartridges in existing shaft element [11, 12, 27].

Smart components or materials can be introduced as complete:

1. Shaft (main)
2. Shafts front—rear
3. Smart one joint
4. Smart both joints.

Smart as contributions in:

5. the intermediate shaft
6. to both shafts (front—Rear)
7. to a joint
8. the both joints.

In shafts can be used mechanical, electro-mechanical, hydromechanical, etc. phase controllers (twisting of the shaft or joints). In the case of shafts the method applies to both parts, in the case of joints—of all components—each alone or in combinations.

The most promising at the moment seems to be the use of smart ceramics type elements for the shafts of their contributions or joints—appeared pilot solutions in Audi and VW vehicles [11].

5 Conclusion

The in the mechanical drive lines structures are introduced radical improvements and radical changes, acting on behalf of the dynamics of vehicle movement. For example—torsional vibration dampers or retarders, assembled in different configurations, using different methods of energy dispersing or recuperation. It shall be also resolved incredibly subtle way—e.g. spline connection modification in several ways. The ability to accomplish tasks and objectives can be achieved with a number of essential tasks. One method uses different regulations yielding through a smart tool. There are some aspects of the use of this tool, and in these particular cases to achieve synchronicity drive.

In the accidental—spatial position of all the shafts, it is possible to maintain synchronicity power transmission. The way is by dynamically adjusting of the phase shift of the joint of the Cardan shaft to the instantaneous position in relative spatial position of the input and output shafts.

A shown method to prevent the emergence of negative dynamic phenomena is through the dynamic adjustment of the phase shift of the Cardan shaft joints to the relative spatial position of the input and output shafts.

Methods involving the elimination of the causes of negative dynamic phenomena, generally require phase adjustment between the joints of the shaft, which must be preceded by specifying of spatial position of the shaft parameters.

Realization of this postulates is possible by a simple algorithm—observation of the state of the spatial position of the shaft and after analysis—to regulate phases.

In the follow-up, diagnostic processes and positions are necessary for the active phase adjustment shaft diagnosed with mileage, algorithms, dynamic, full and partial results positions of the shaft and the dynamic effects of machinery or vehicle temporary state planning. The aim is usually to control the prediction of the developing destructive process to avoid undesirable conditions or operations border limits. It is necessary to build diagnose for shaft during the operation change the stiffness of the support, suppression of internal rigidity of components, and even weight. CAD methods and broad CAE (E) will be here almost optimal. These components also exhibit different preliminary characteristics. Diagnostic method should allow the assessment of the spatial position of the shaft, the shaft dynamics, and ultimate phase adjustment shaft.

Elements of the smart systems may be introduced as part of the integral of the Cardan shaft or as inserts to existing shaft elements. Components, or materials intelligent (smart) can be introduced as a complete shaft, shaft components, joints, or contributions to them, or external phase controllers with different technical structure and operational rules.

References

1. Anantapal, N.: The application concept of universal joint to make limited slip differential, center differential and clutch. SAE Technical Paper (2007). https://doi.org/10.4271/2007-01-0662
2. Avallone E.A., Baumeister III T., Sadegh A.M.: Marks' Standard Handbook for Mechanical Engineers. The McGraw-Hill Companies (2007)
3. Cornay, P., Britton, J.: New variations of the cardan concept increase universal joint performance. SAE Technical Paper (1991). https://doi.org/10.4271/911777
4. Fischer, I.: Effect of mounting errors on cardan-type universal joint kinematics. SAE Technical Paper (1988). https://doi.org/10.4271/880483
5. Flesch, G., Jaskulski, L.: The application of design of experiments as a support tool for a drive shaft retention system development. SAE Technical Paper (2005)
6. Garnham, J.: Constant velocity joint material performance criteria. SAE Technical Paper (1996). https://doi.org/10.4271/960572
7. http://www.uni.edu/~rao
8. Iqbal, J., Qatu, M.: Vibration analysis of a three-piece automotive shaft. SAE Technical Paper (2009). https://doi.org/10.4271/2009-01-2067
9. Johnson, R.: High speed fixed constant velocity joint for automotive driveshaft applications. SAE Technical Paper (1998). https://doi.org/10.4271/980834
10. Kulczyk I.: Wały napędowe, półosie, przeguby wałów i półosi. Bydgoszcz (2011)
11. Kunze, H., et al.: Vibration Reduction on Automotive Shafts Using Piezoceramics. Fraunhofer-Institute for Machine Tools and Forming Technology, Dresden/Germany), Volkswagen AG, Vibration-X, Winchester, MA, USA, Fraunhofer Institut Werkzeugmaschinen und Umformtechnik
12. Loh et al.: United States Patent US006752425B2. Semi-active control of automotive steering system vibration magneto_rheological damping, June 22, 2004
13. Lee, C., Polycarpou, A.: Experimental investigation of tripod constant velocity (CV) joint friction. SAE Technical Paper (2006). https://doi.org/10.4271/2006-01-0582
14. Maciag, W., Mushenski, M.: New bearing design concept an innovative, U.S. army, design concept for tactical vehicle bearings and universal joints. SAE Technical Paper (1997). https://doi.org/10.4271/973178

15. Magirius, S., Booker, D.: High speed constant velocity joints for car and light truck driveshafts. SAE Technical Paper (1995). https://doi.org/10.4271/950891
16. Norton, R.L.: Machine Design: An Integrated Approach, 3rd edn. Prentice Hall (2006)
17. Park, B., Stühler, W.: Dynamic Behaviour of a Universal Joints. SAE Technical Paper (1991)
18. Parmley, R.O.: Illustrated Sourcebook of Mech. McGraw-Hill, Components (2000)
19. Przystupa, F.W.: Diagnozer w systemie technicznym. Publishing House Oficyna Wydawnicza Politechniki Wroclawskiej, Wroclaw (2010)
20. Ramachandra, S.: Theoretical analysis for practical design of universal joint trunnion bearings. SAE Technical Paper (1986). https://doi.org/10.4271/860387
21. Sakakibara, M.: Design CAE approach applied to drive train components. SAE Technical Paper (1995). https://doi.org/10.4271/950902
22. Shigley, J.E., Mischke, C.R., Brown Jr., T.H.: Standard Handbook of Machine Design. The McGraw-Hill Companies (2004)
23. Tsuda, M., Kojima, H., Arita, M.: The development on cold forging technique to form a component of the constant velocity joint. SAE Technical Paper (1986)
24. Ullman, D.: The Mechanical Design Process, 3rd edn. McGraw-Hill (2002)
25. Vedam, K., et al.: Analysis of an automotive driveline with cardan universal joints. SAE Technical Paper (1995). https://doi.org/10.4271/950895
26. Yamamoto, T., Matsuda, T., Okano, N.: Efficiency of constant velocity universal joints. SAE Technical Paper (1993). https://doi.org/10.4271/930906
27. Z Car, Diytrade, Pelican Parts, Trico Driveshaft Company, Atlantic
28. Gubran, H.B.H., Gupta, K.: Design optimization of automotive propeller shafts. J Vib. Eng. Technol. **2**(1) (2014). ©Krishtel eMaging Solutions Pvt. Ltd.
29. Bankar, H., et al.: Material optimization and weight reduction of drive shaft using composite material. IOSR J. Mech. Civil Eng. (IOSR-JMCE) **10**(1). e-ISSN: 2278-1684, p-ISSN: 2320-334X (2013)
30. Sheikh, S.M.A.: Analysis of universal coupling under different torque condition. Int. J. Eng. Sci. Adv. Technol. [IJESAT] **2**(3), 690–694. Available online at http://www.ijesat.org (2012)
31. Yu, R.S., et al.: Numerical investigation on the dynamic behaviour of advanced ceramics. Eng. Fract. Mech. **71**, 897–911 (2004)
32. Khoshravan, M.R., Paykani, A.: Design of a composite drive shaft and its coupling for automotive application. J. Appl. Res. Technol. **10** (2012)
33. Slocum, A.H.: ME 2.075 Course Notes. MIT (2001)
34. Weck, M., Brecher, C.: Werkzeugmaschinen (Band 2). Springer-Verlag (2005)

Problems of Heat Transfer Modelling in Cooling Systems of Earthmoving Machinery

Stanisław Kwaśniowski

Abstract In working machinery with an internal combustion drive, about 60% of the energy released in the combustion process is converted into heat and is discharged directly into the environment (with exhaust gases) or indirectly through the cooling system of the power unit [8–10]. The efficiency of the cooling system and the temperature level of the media interposing the heat transfer determine the correct operation of the machine. These parameters are particularly important in extreme operating conditions, e.g. starting the machine in severe winter conditions or operating the machine in a tropical climate. The answer to relevant questions about the thermal state of individual units can be obtained through simulation studies of heat transfer dynamics on the experimentally verified mathematical models. The paper presents the idea of using calculation models in the form of systems of ordinary, non-linear differential equations, in which the MATLAB–SIMULINK package was used.

Keywords Industrial machines and vehicles · Cooling systems · Dynamics of heat exchange · Temperature simulations in complex structures

1 Introduction

A crucial technical problem that arises in internal combustion engines is thermal loads. This issue becomes more complex when we consider the thermal effects of other units combined with the engine. The engine together with the combined units (Fig. 1) creates a complex heat transfer system.

This issue becomes more complex when we consider the thermal effects of other units combined with the engine. The engine together with the combined units (Fig. 1) creates a complex heat transfer system.

The thermal states of all elements of the system are mutually coupled, which is why the presented system of mutual interactions results in the need for it to be considered as a whole, together with the influence of the environment. The range of

---

S. Kwaśniowski (✉)
Faculty of Mechanical Engineering, Wroclaw University of Science and Technology, ul. Wybrzeze Wyspianskiego 27, 50-370 Wroclaw, Poland
e-mail: stanislaw.kwasniowski@pwr.edu.pl

M. Sokolski (ed.), *Mining Machines and Earth-Moving Equipment*,
https://doi.org/10.1007/978-3-030-25478-0_8

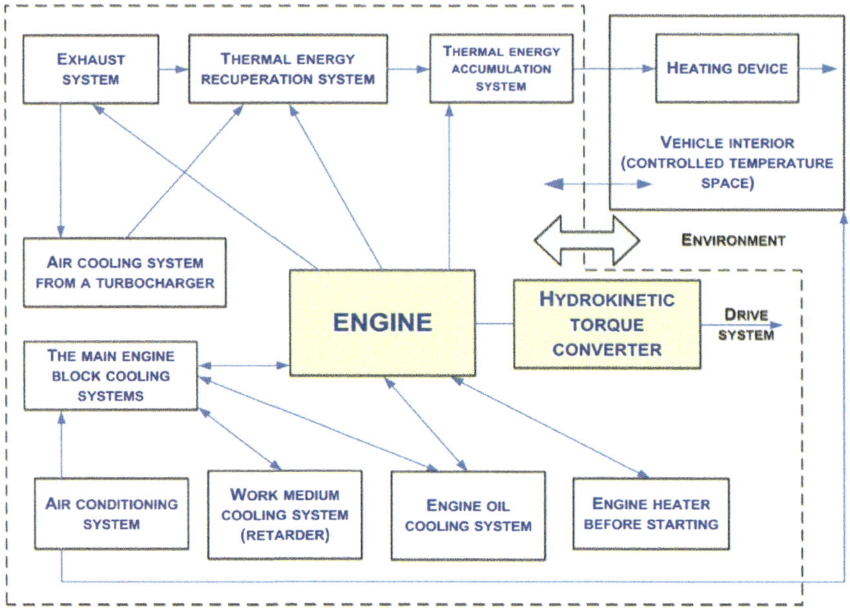

Fig. 1 Functional structure of thermal devices combined with the combustion power unit. *Source* [1, 10]

heat transfer systems in vehicles with an internal combustion engine is wide. The selection of appropriate heat exchangers, their capacity, and thermal efficiency is a very important design issue, as it affects the thermal state of the entire system. When configuring the cooling system, it is important to remember that the load of the power unit, and thus of the entire cooling system, is determined by:

- climatic conditions,
- terrain and road conditions,
- useful loads.

On the one hand, the cooling system should provide us with proper engine operation in extreme ambient temperatures and loads of the power unit, but on the other hand this system cannot be oversized because it involves energy and material consumption in the production process, unnecessary transport of additional weight during the operational period and additional unjustified costs. Poor configuration of the cooler system can cause the cooling effects of one of them to be voided by another. The aim is to obtain an optimal structure, i.e. one that would have sufficient capacity for the assumed operating conditions at the lowest possible curb weight and capacity. The mass and geometric parameters of the structure determine the area of safe operation of the cooling system with no fear of overheating the coolants. These fluids have a considerable impact on the efficiency of cooling, e.g., replacing the water in the engine system with glycol-based liquid or another reduces the efficiency of the cooling system from a few to a dozen percent, instead of enabling operation in negative or elevated ambient temperatures.

To sum up, it is important to maintain in the system such as temperatures, which will ensure proper operation of the engine. This can be achieved by the appropriate selection of the cooling system.

The aim of this work is to present a calculation model that facilitates the assessment of heat flux and temperature levels in individual elements of the extended cooling system of the power unit, on the example of the Ł240 loader with the 6C107 engine.

2 Structure of Cooling System of Ł240 Loader's Power Unit

The cooling system of the described machine (Fig. 2) consists of the main cooler (3)—water–air and engine oil cooler (10)—cooled by water returning from the main cooler.

Thanks to the use of a hydrodynamic transformer, it is possible to start a heavy machine almost without load, stepwise. However, not all power transmitted from the internal combustion engine is converted into work. The rest of the energy that has not been converted into mechanical energy is converted into heat and goes to the

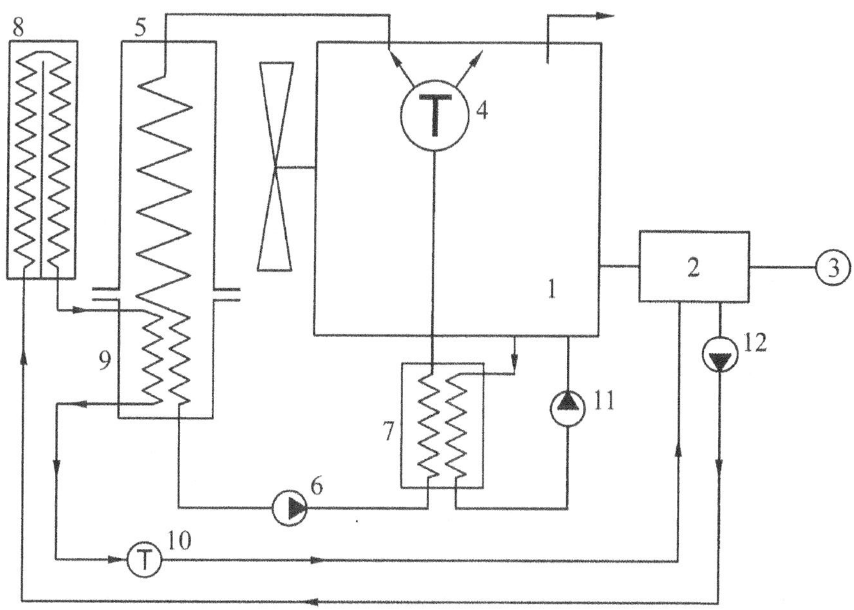

Fig. 2 Structure of the Ł240 loader's cooling system [7]: 1—engine, 2—drive system, 3—water–air cooler, 4—thermostat of the cooling system of the engine block, 5—main cooler of the engine, 6—coolant circulating pump, 7—transmission oil cooler, 8—integrated transmission oil cooler, 9—thermostat, 10—engine oil cooler, 11—hydrodynamic transformer

exhaust and cooling system. In particularly tough operating conditions at full slip on the transformer, there are periods in which almost the entire flux of mechanical energy is converted into heat. In this situation, the cooling system should have a thermal efficiency almost twice as high as the rated power of the engine. That is why it is so important to use an additional cooling system. The transmission oil is cooled in an oil cooler (7), usually placed in front of the engine, or in a liquid–liquid cooler (8), placed in the lower tank of the main water cooler. The oil circulation is forced by means of a special oil pump.

One may wonder if the system described is efficient—its serial design in the case of a cascade of coolers (3), (8) and (10) may cause that with an increased temperature of water at the input to the radiator (10) its efficiency will be low. However, taking into account the recommended level of engine oil temperature, excessive cooling of the engine oil would be undesirable from the point of view of engine durability. The answer to this question can be obtained by creating a heat balance of the engine under the agreed operating conditions. The design and flow features of individual heat exchangers determine the temperature level of the coolant.

3 Rules for Computations on Heat Exchangers

The calculation model of the cooling system is based on the relationships describing the heat exchange in individual highlighted elements and units. The heat fluxes obtained in this way according to Newton's law [6, 8, 12] are expressed by the element-specific temperatures variable in time. The heat transfer in a dedicated heat exchanger is described by the following balance equation:

$$Q = m_1 \cdot c_1 \cdot (T_{11} - T_{21}) = m_w \cdot c_w \cdot (T_{2w} - T_{1w}) = k \cdot F \cdot \left(\bar{T}_w - \bar{T}_1\right) \qquad (1)$$

where:

m mass medium's flow rate,
c specific heat of the coolant,
k coefficient of heat transfer through walls of the cooler,
F radiator's heat transfer surface,
T temperature,
$\bar{T}$ average temperature of medium in the exchanger.

Indexes:

1 for input,
2 for output,
l for air,
w for water.

The k heat transfer coefficient in Eq. (1) includes the thermo-physical parameters that determine the heat transfer between the media (Peclet coefficient) [8, 9]:

$$k = f(\alpha_1, \alpha_w, \lambda, s, \varepsilon, \delta) \tag{2}$$

where

α heat transfer coefficient,
λ heat conductivity,
s thickness of walls of cooler pipes,
ε ribbing efficiency,
δ cooler's ribbing coefficient.

Indexes as in dependency (1).

The values of the heat transfer coefficients on the air α_1 and waterside α_w depend on the parameters of thermal and flow heat transfer media. These dependencies are described by characteristic numbers [10]:

$$\alpha = Nu \cdot \lambda/d_n \tag{3}$$

$$\mathrm{Nu} = c \cdot \mathrm{Re}^A \cdot \mathrm{Pr}^B \tag{4}$$

where

Nu Nusselt number,
Re Reynolds number,
Pr Prandtl number,
d_n hydraulic diameter,
λ thermal conductivity of the medium.

The values of coefficients *A*, *B*, *C* are determined empirically for individual typical cases of exchanger designs [5, 8, 10]. Due to the non-linear nature of the dependence (3) and (4), Eq. (1) has the character of an implicit equation, and its solution can be obtained with a numerical method. The relationships describing heat transfer in oil–air, oil–water and air–air exchangers have a similar nature.

The adopted method of calculation (MBE elementary balance method) consists in dividing the heat exchanger's core into elementary sections and describing the heat transfer between media within a given section with coordinates (*i*, *j*, *k*), where *i*, *j*, *k* are section numbers along three Cartesian coordinates (*x*, *y*, *z*). Thus, the heat transferred across the Q_{ch} cooler is the sum of heat fluxes Q_{ijk} in individual elements:

$$Q_{ch} = \sum_i \sum_j \sum_k Q_{ijk} \tag{5}$$

It is like a triple integral $\int_a \int_b \int_c f \mathrm{d}x\mathrm{d}y\mathrm{d}z$.

Examples of diving cores of typical exchangers found in cooling systems of power units are shown in Fig. 3 [8].

When constructing calculation models [8, 10] of heat exchangers, the following simplifying assumptions are adopted:

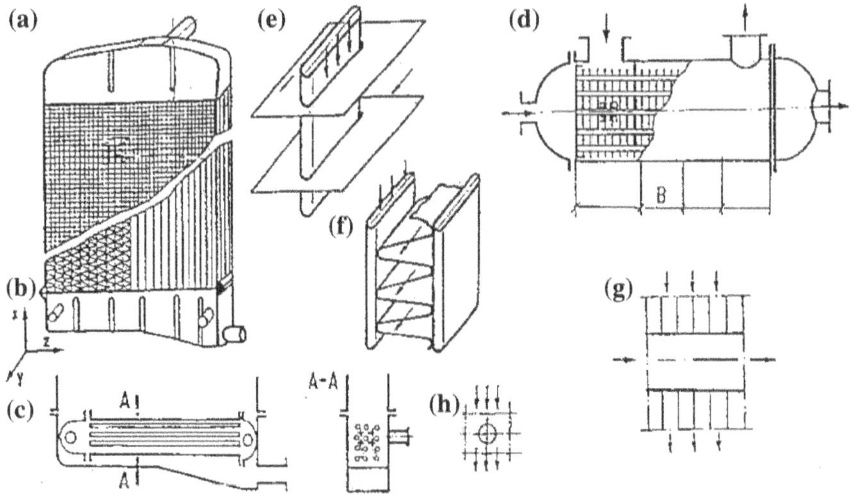

Fig. 3 Examples of cooler constructions: **a** water–air cooler with plate core, **b** water–air cooler with RTST steel core, **c** cooler of transmission oil (water–oil), **d** cooler engine oil (water–oil), **e** elementary section of an a-type cooler, **f** elementary section of a b-type cooler, **g** elementary section of a c-type cooler, **h** elementary section of a d-type cooler

1. Flows of heat transfer media are uniform throughout the cross-section of the exchanger.
2. The exchanger core consists of tube columns and ribbing—heat transfer in each column is uniform, i.e. local heat transfer interferences near the housing and interferences resulting from uneven flow of media.
3. Heat transfer is assumed in two directions: in the tube axis and at the thickness of the exchanger (between rows of tubes), whereas due to parallel nature of the flow, heat exchange between the columns is omitted.
4. Final temperature of heat exchange media in the extracted element is the initial temperature for the adjacent element.
5. Specific heat of the media flowing through the exchanger in the range of the temperature considered is unchanged; however, the change in viscosity and thermal conductivity as a function of temperature is taken into account.
6. With regard to heat exchangers in which one of the media is atmospheric air, it is assumed that for the adjustable tube arrangement, the "k" heat transfer coefficient has a value higher by 10%—this results from the experimental verification of the calculation model.
7. Thermal losses on the pipes connecting the exchangers are omitted.

The described calculation principles of individual exchangers allow to calculate the temperature distribution of heat carriers in the analyzed heat exchangers. Due to the non-linearities of changes in thermo-physical properties of heat carriers (thermal conductivity, specific heat, dynamic viscosity) and mutual effects, the thermal balance equation of the entire cooling system is transcendental equation and can be solved only numerically. For practical purposes, the authors used the Gauss–Seidel method to solve it.

4 Modelling of Heat Transfer Dynamics

4.1 Theoretical Basics

From the point of view of the global assessment of the engine's operating states, it may be useful to evaluate heat fluxes and temperatures in elements and units, the state of which is described using concentrated parameters variable in time, e.g. average temperature or global heat flux exchanged with the environment by an element or a unit. The structure of the system understood in this way can be presented graphically, which is helpful in the construction of mathematical models.

The basis for the design of complex structures of heat exchange are m mass and *p* flow elements (Fig. 4), where the q_{ij} symbol indicates heat fluxes exchanged between *i* and *j* systems, and the *q* symbol and the flow heat accumulated in the *i* element has been indicated [6, 11].

In the first case (*m*), the relationship describing this structure will take the form:

$$q_i = q_{i-1,i} - q_{i,i+1} \tag{6}$$

In the differential notation:

$$V_i \cdot \rho_i \cdot c_p \cdot \dot{T}_i = q_{i-1,i} - F_{i,i+1} \cdot \alpha_{i,i+1} \cdot (T_i - T_{i+1}) \tag{7}$$

where:

$q_{i-1,i}$ thermal energy directed to the system,
V_i volume of *i* element,

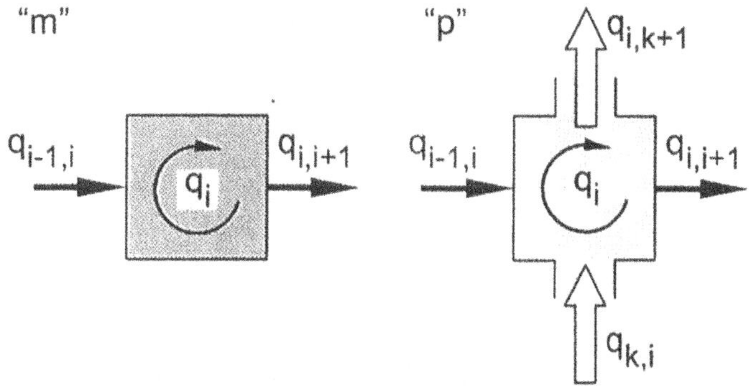

Fig. 4 Basic elements *m*—mass element, *p*—flow element

ρ_i density of i element,
$F_{i,i+1}$ the surface of heat transfer between i and $i + 1$ elements,
$\alpha_{i,i+1}$ heat transfer coefficient between i and $i + 1$ elements.

In the second case (p), the relationship will take the following form:

$$q_i = q_{i-1,i} - q_{i,i+1} + q_{k,i} - q_{i,k+1} \tag{8}$$

In the differential notation:

$$V_i \cdot \rho_i \cdot c_p \cdot \dot{T}_i = q_{i-1,i} - F_{i,i+1} \cdot \alpha_{i,i+1} \cdot (T_i - T_{i+1}) + \dot{V}_{k,i} \cdot \rho_{k,i} \cdot T_k - \dot{V}_{i,k+1} \cdot \rho_{i,k+1} \cdot T_i \tag{9}$$

The symbols are analogous to (7).

On the basis of m mass and p flow elements, elements of $m + p$, $m + p + m$, $p + m + p$ types can be built (Fig. 5), which correspond to the structure of pipes with the heat carrier, heat exchangers, etc.

As an example of the discussed principle of building models of heat exchange dynamics, a model of an engine cooling system equipped with a water cooler, a transmission oil cooler and an engine oil cooler will be presented. The structure of this system in block form will be presented later in the article.

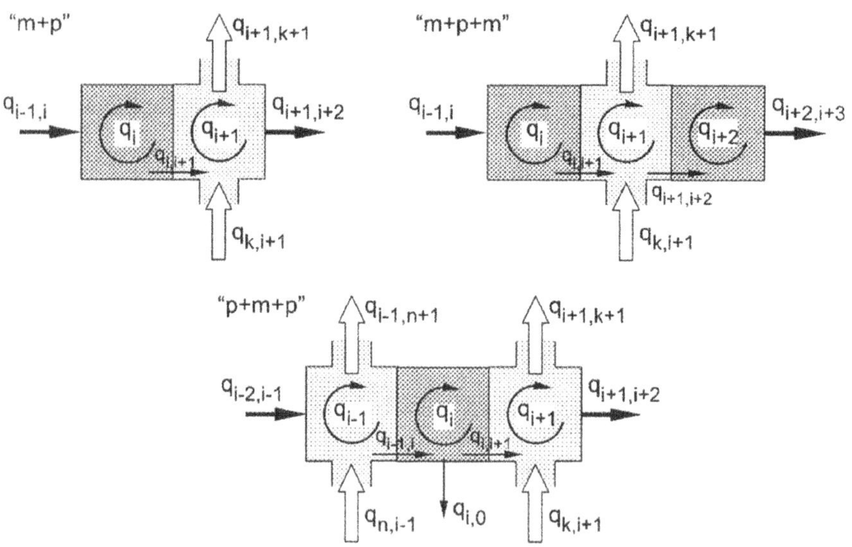

Fig. 5 Basic structures present in cooling systems of internal combustion engines and heat exchange systems in vehicles

4.2 Model of the Engine with a Water–Air Cooler and Engine Oil Cooler

The engine model is shown in Fig. 6.

The heat exchange in the model thus adopted can be illustrated by the following block diagram of heat exchange (Fig. 7).

The presented model of heat exchange in the engine is described by ten balance equations:

$$\begin{aligned} & q_1 = q_{dost} - q_{1,2} - q_{1,6} - q_{1,5};\ q_2 = q_{1,2} - q_{2,3} - q_{3,2} \\ & q_3 = q_{2,3} - q_{3,4} - q_{3,2} - q_{3,14};\ q_4 = q_{3,4} - q_{4,0} - q_{5,4} \\ & q_5 = q_{5,6} - q_{5,0} - q_{5,4} - q_{5,1};\ q_6 = q_{1,6} - q_{5,6} - q_{6,8} - q_{6,14} \\ & q_8 = q_{6,8} - q_{8,12} - q_{8,14};\ q_{12} = q_{8,12} - q_{12,13} \\ & q_{13} = q_{12,13} - q_{0,13} - q_{13,0};\ q_{14} = q_{3,14} - q_{8,14} - q_{6,14} \end{aligned} \tag{10}$$

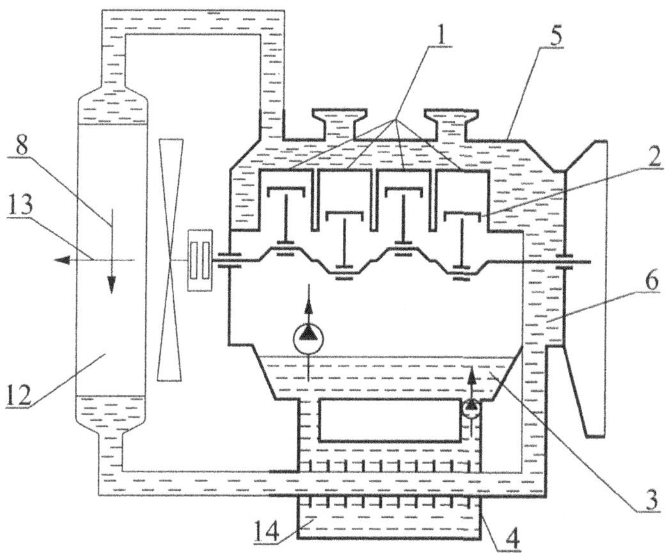

Fig. 6 The physical model of 6C107 combustion engine: 1—hot engine components (surrounding the combustion chamber), 2—oil in the engine block, 3—oil in the oil pan, 4—housing of engine oil cooler, 5—external engine walls, 6—water in the engine block, 8—water in main cooler, 12—cooler core, 13—radiator air space (air mass filling the cooler core), 14—oil in the engine oil cooler

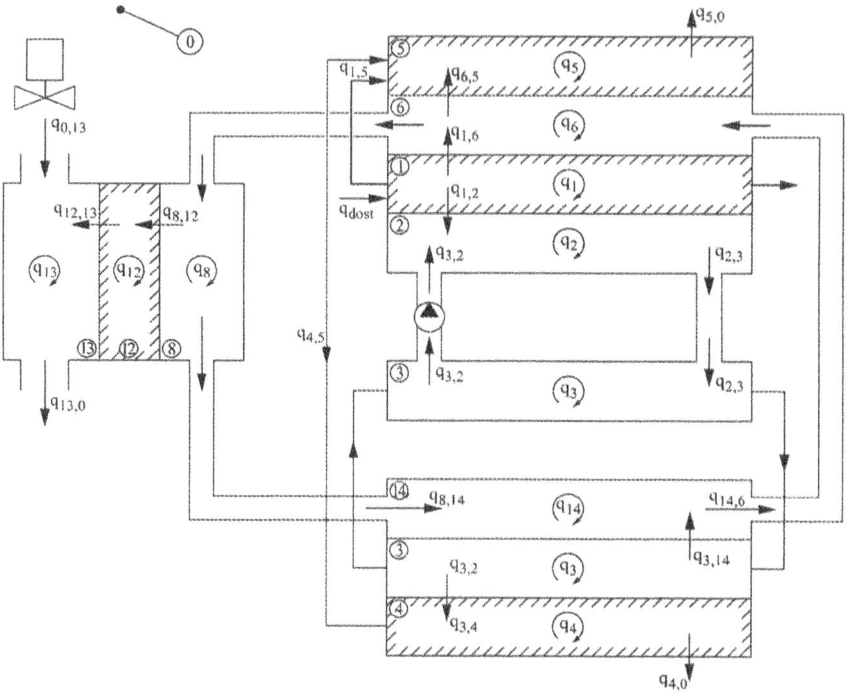

Fig. 7 Model diagram of the engine with a water–air cooler and engine oil cooler: 0—environment, 1—hot engine components (surrounding the engine chamber), 2—oil in the engine block, 3—oil in the oil pan, 4—housing of the oil cooler, 5—external engine walls, 6—water jacket in the engine block, 8—water space of the cooler, 12—mass of the cooler, 13—air space of the cooler, 14—oil in the engine oil cooler

The presented physical model was described using ten differential equations. Based on the analysis of the design of the 6C107 engine and on-site verification in the Laboratory of Internal Combustion Engines of Wroclaw University of Technology [1, 7], the coefficients describing the thermophysical properties of the construction elements and the properties of the heat carriers have been determined. Such a mathematical model was used for simulation testing [1, 8]. Examples of simulation results on temperatures in the engine are shown in Fig. 8.

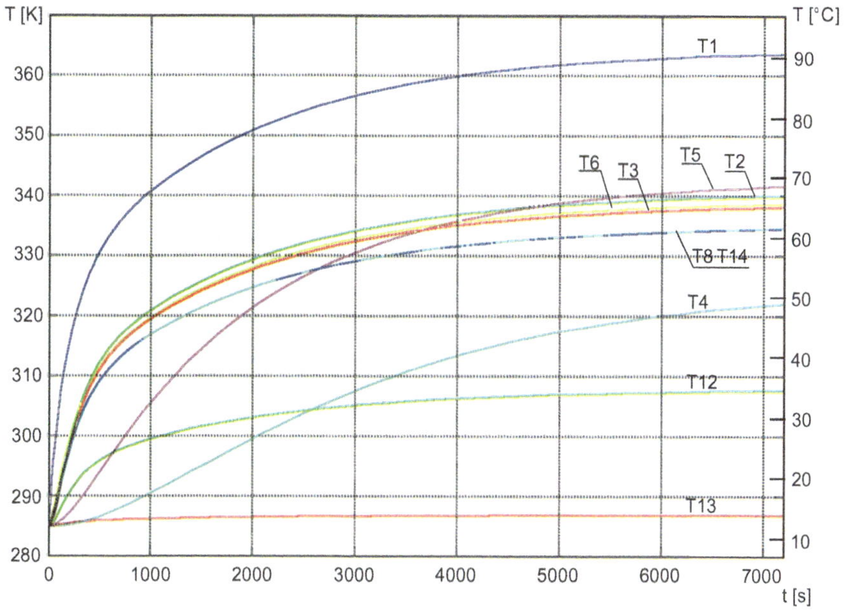

Fig. 8 Results of engine speeding-up simulation with the water–air cooler and the engine oil cooler: T1—temperature of hot engine elements, T2—oil temperature in the engine block, T3—oil temperature in the oil pan, T4—temperature of the housing of oil cooler, T5—temperature of the external walls of the body, T6—water temperature in the engine block, T8—water temperature in the cooler, T12—temperature of the cooler's core, T13—air temperature in the cooler's air space, T14—water temperature in the engine oil cooler

5 Simulation Tests on Heat Transfer Dynamics in Specific Operating Cycle of the Machine

5.1 Characteristics of the Calculation Model

The tested object was the cooling system of the Ł240 loader (Fig. 9) from FADROMA— Wrocław.

There were two sources of heat in the machine: the engine and the torque converter. The cooling system consisted of four coolers: the main water–air cooler, the oil–air transmission oil cooler, the additional transmission oil cooler and the oil–water engine oil cooler. As a basis for the calculation model [5, 8], the 6C107 engine model was experimentally verified.

This system in block form is shown in Fig. 10.

Fig. 9 Ł240 loader

5.2 *Characteristics of Machine Loading Conditions in a Specific Operating Cycle*

In this analysis, the work of the loader was assumed in a typical operating cycle of loading bulk solid onto a vehicle (car or wagon).

In this cycle (Fig. 11), eight characteristic phases can be distinguished.

On the basis of research conducted by Professor Pieczonka and his colleagues [2–4] from the Institute of Machine Design and Operation of the Wrocław University of Technology, the following parameters characterizing the adopted operating cycle were concluded (Table 1).

Knowing the power values, the efficiency values of the torque converter can be determined with the close approximation, and the heat released from the converter can then be calculated:

$$q_{\mathrm{dost2}} = N \cdot (1 - \eta) \tag{11}$$

where

q_{dost2} thermal power emitted from the torque converter,
N engine power,
η efficiency of torque converter.

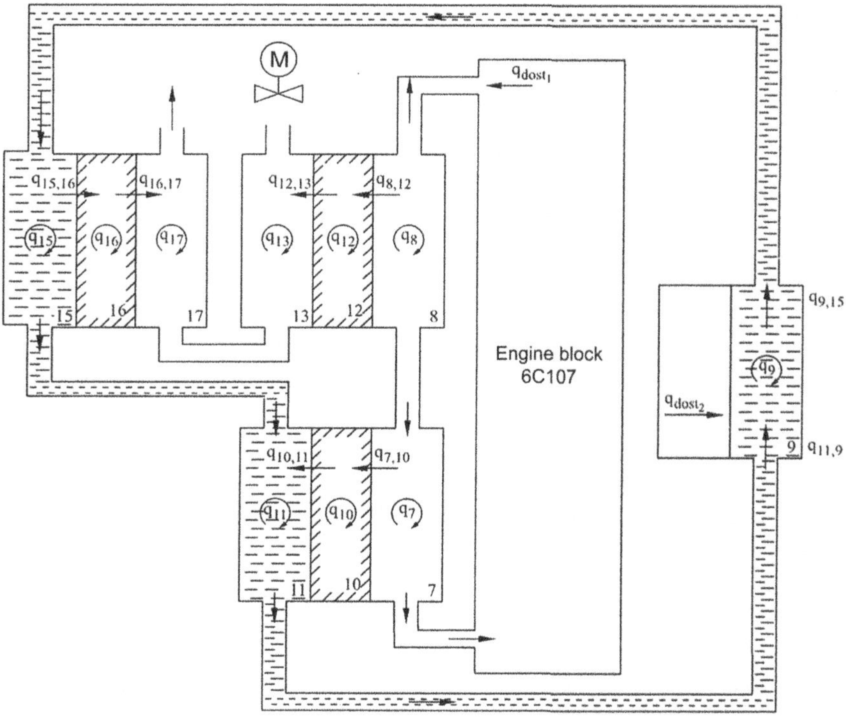

Fig. 10 Block diagram of the cooling system of the Ł240 loader's power unit: 0—environment, 1—hot engine components, 2—oil in the engine block, 3—oil in the oil pan, 4—oil pan, 5—external engine walls, 6—water jacket in the engine block, 7—water space of water-oil cooler, 8—water space of the cooler, 9—workspace of the torque converter, 10—weight of water-oil cooler, 11—oil space of water-oil cooler, 12—cooler's mass, 13—air space of the cooler, 14—cooler of engine oil, 15—oil space of the oil–air cooler. 16—mass of the oil–air cooler, 17—air space of the oil–air cooler

Fig. 11 Loader operating cycle: 1—approach to the piling, 2—lowering the scoop, 3—filling the scoop with slight raising, 4—backing the loader, 5—approach to the conveyance mean, 6—lifting the scoop, 7—dumping, 8—backing the machine and lowering the scoop

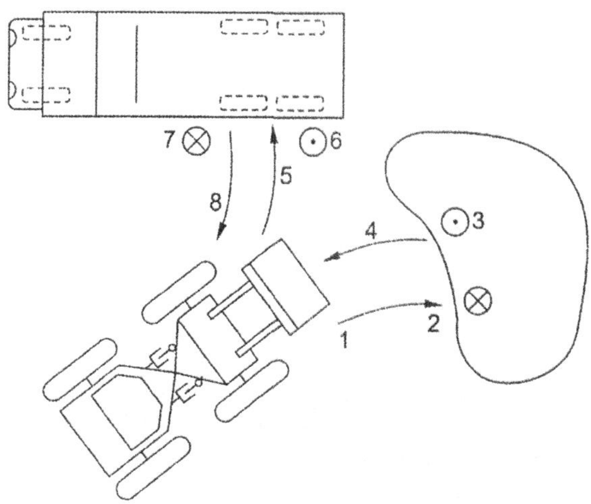

Table 1 Parameters characterizing the individual phases of the loader's movement

Phase of motion	t (s)	N (kW)	n (rpm)	η (%)	q_{dost2} (kW)
1	5	30.5	1500	80	6.1
2	3	10.0	700	20	8.0
3	6	91.5	2400	10	82.35
4	6	35.5	2000	80	7.1
5	8	35.5	2000	80	7.1
6	4	61.0	1500	60	24.4
7	3	20.5	700	20	16.4
8	5	30.5	1500	80	6.1

where

t duration of the movement phase,
n average engine power in a specific phase of movement,
N engine rpm,
η converter efficiency,
q_{dost2} thermal power emitted by the torque converter to the cooling system.

The time of one full cycle is 40 (s).
The average power value: 39.38 (kW).
The average value of the torque converter efficiency: 53.75 (%),
The average value of thermal power of the torque converter: 19.70 (kW).

The values of heat emitted by the torque converter in the individual phases of the loader's operating cycle for the cooling system are presented below (Fig. 12).

Examples of temperature curves in the cooling system of the tested loader are shown in Figs. 13 and 14.

The variable cycle of the machine's operation induces typical temperature pulsations, which in a fixed operating state (powered-up system) range from a few to almost 20 K. These pulsations can cause momentary boiling of the coolant. The level of pulsation and the average values of coolants also depend on the ambient temperature in which the machine operates.

5.3 Results of Simulation Tests

Based on the developed calculation model, simulation tests were carried out for temperatures of water, transmission oil, and engine oil.

These are two variable parameters on which the thermal state of the cooling system depends: the ambient temperature at which the machine operates and the efficiency of the torque converter. The second parameter has already been specified in the loader's operating cycle.

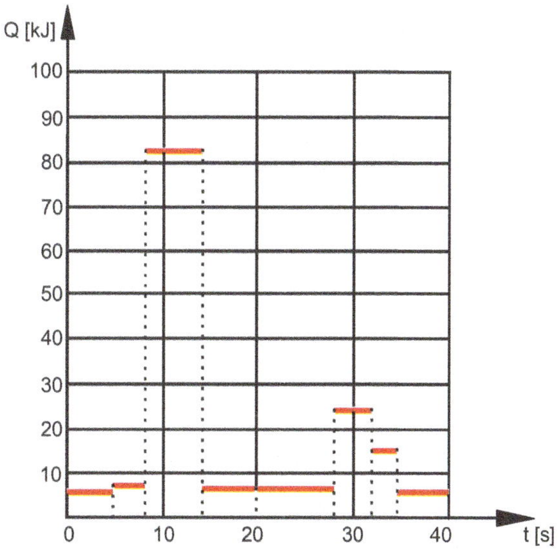

Fig. 12 Dependence of heat emitted from the torque converter as a function of time

The tests were carried out for various five ambient temperatures. The results of calculations are presented in Table 2.

As can be seen in the simulation process and Fig. 15, the correct operation of the cooling system of the machine may occur at ambient temperatures not exceeding 298 K (25 °C). Above this limit, the operation takes place in a dangerous area. Water reaches its boiling point at the temperature a few degrees higher. The value of water temperature above 100 °C is the calculative temperature. However, the main limitation of the safe working area of the system is the temperature of the transmission oil. The permissible operating temperature of transmission oil is 398 K (125 °C) and for engine oil it is 408 K (135 °C).

Taking into account the upper-temperature limit of the transmission oil, the system must not exceed 295 K (22 °C) of ambient temperature. For the lower limit, this value is 315 K (42 °C). This indicates that the efficiency of the water cooling system should be increased, and the cooling of transmission oil should be streamlined. Otherwise, it will be necessary to stop operation in order to cool down the engine. Because the obtained results indicate that the machine is not capable of operation at higher ambient temperatures, simulation tests of the system were carried out at water flow in the main cooling system increased by 50 and 100%.

In graphical form, the results of calculations are illustrated in Fig. 15.

Temperature values of individual media at 298 K (25 °C) ambient temperature in a specific operating cycle are listed in Table 3.

As the obtained results indicate that the machine is not capable of operation at higher ambient temperatures, system simulation tests were carried out with increased water flows in the main cooling system.

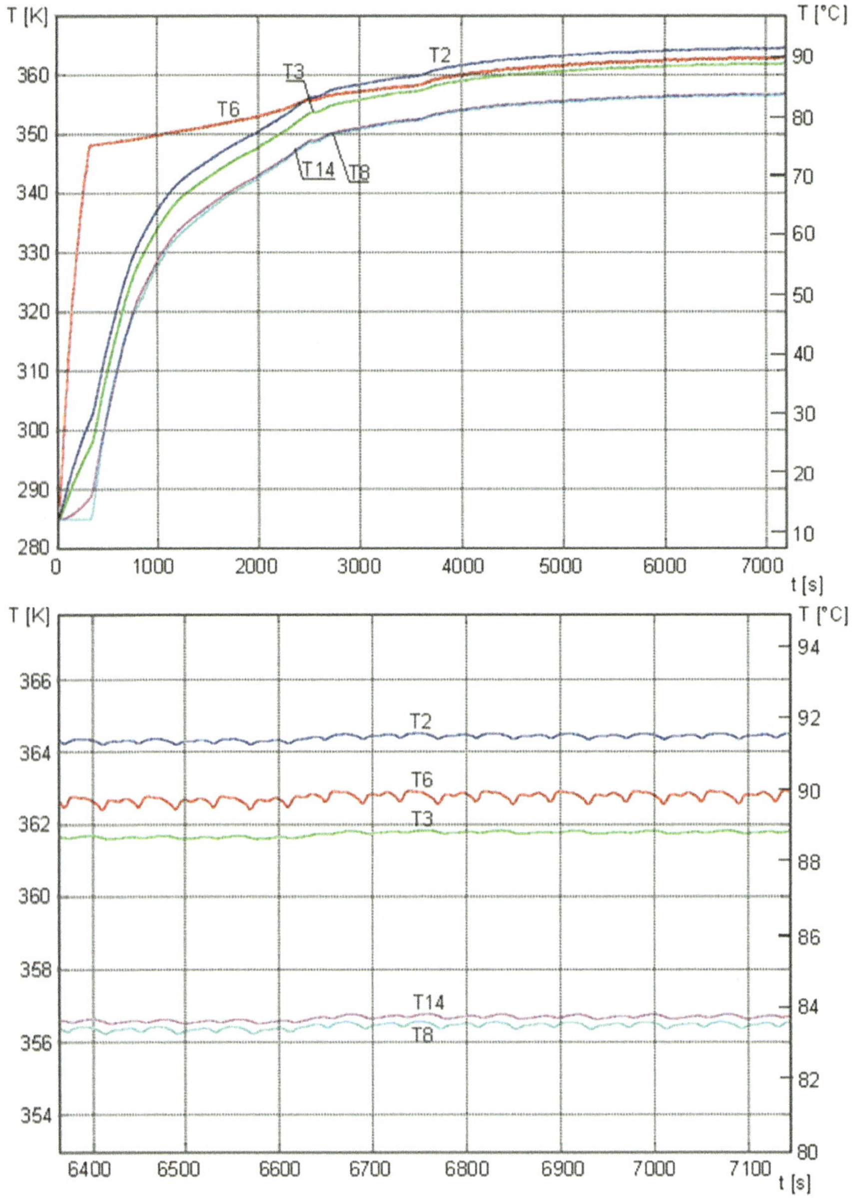

Fig. 13 Simulation results of the cooling system of the loader with thermostat and variable load: T2—oil temperature in the engine block, T3—oil temperature in the oil pan, T6—water temperature in the engine block, T8—water temperature in the water–air cooler, T14—water temperature in the engine oil cooler

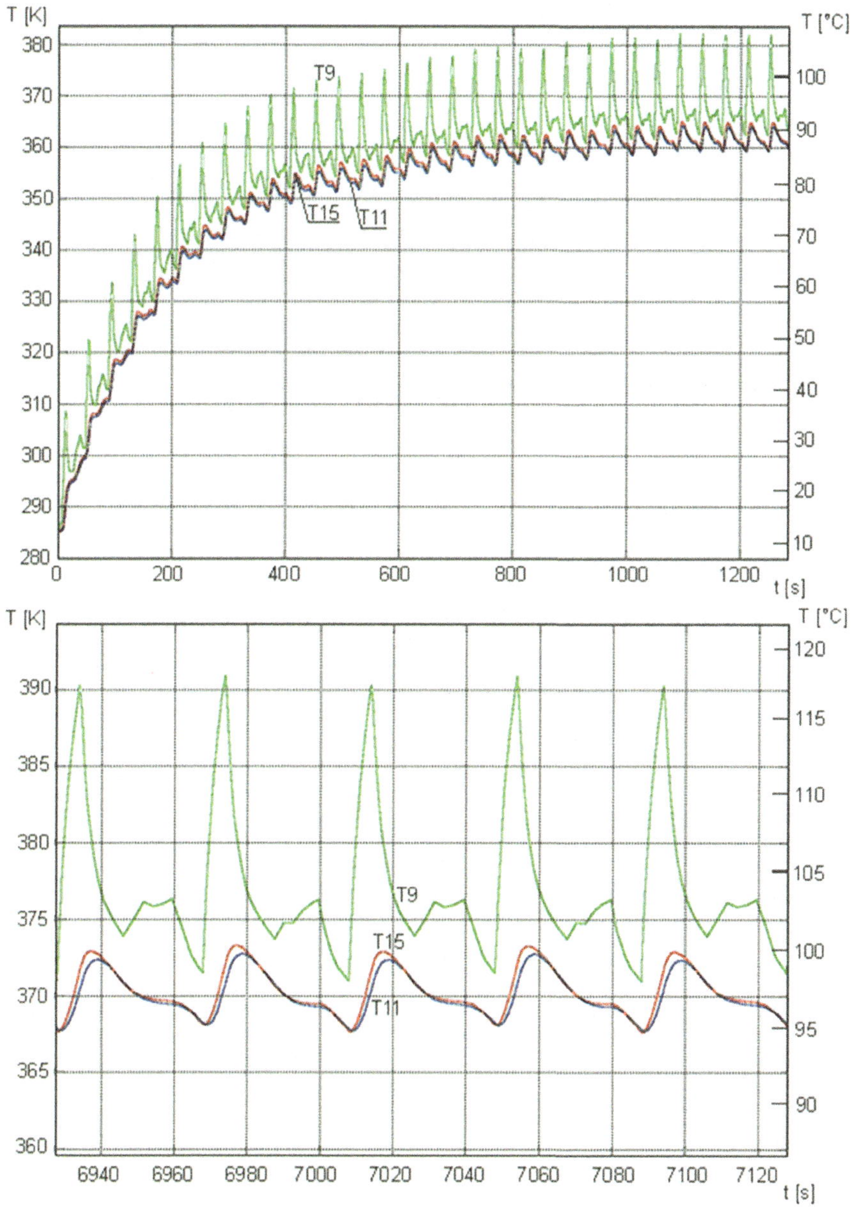

Fig. 14 Transmission oil's temperature diagram in different parts of the cooling system: T9—temperature emitted by the torque converter, T11—oil temperature in the water-oil cooler, T15—oil temperature in the oil–air cooler

Table 2 Media temperature values in the system depending on the ambient temperature

Ambient temperature (°C)	Temperature of water (°C)	Temperature of engine oil (°C)	Temperature of gear oil (°C)
−30.0	76.5	54.0	54.0–74.0
−10.0	78.5	70.5	74.0–94.0
5.0	82.5	82.5	88.0–108.0
20.0	94.0	95.5	103.5–123.5
40.0	116.0	117.5	123.0–143.0

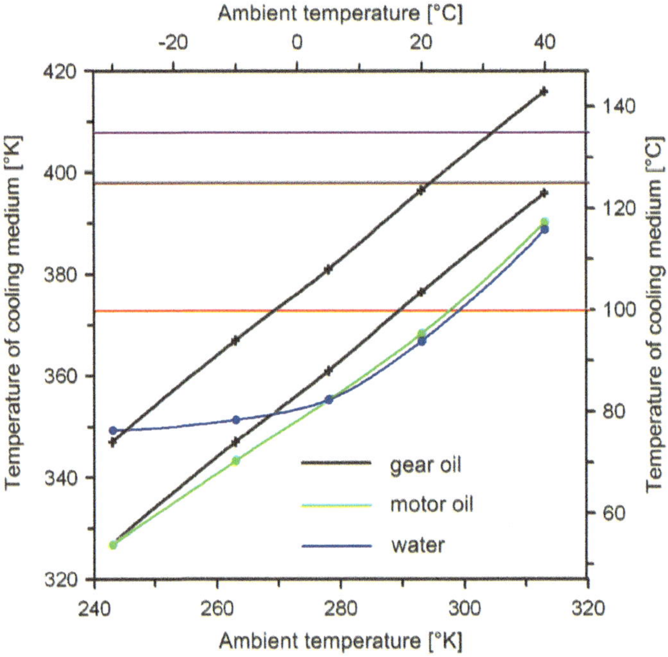

Fig. 15 Temperature dependency of water, engine oil and transmission oil in the cooling system from ambient temperature (the red line indicates the water's boiling point, the brown line indicates the permissible operating temperature of the transmission oil (398 K), the violet line indicates the permissible operating temperature of the engine oil (408 K))

Table 3 Values of media temperatures in the system depending on the level of water flow

Temperature	Normal water flow	Water flow increased by 50%	Water flow increased by 100%
T9 (°C)	110.0–130.0	102.0–122.0	99.5–119.5
T2 (°C)	104.0	81.0	72.1
T6 (°C)	102.2	82.2	78.3
T3 (°C)	101.2	79.0	70.1
T14 (°C)	96.3	76.5	68.2

The calculations show that increasing the water flow in the cooling system significantly reduces the temperatures of other media and allows the machine to operate at elevated ambient temperatures.

6 Summary

The aim of the paper was to present tools useful in the analysis of heat transfer dynamics in an extensive heat exchange system.

An example of such a structure was the cooling system of the Ł240 loader. The basis of the calculation model was a system of differential and non-linear equations. This system was solved using the MATLAB-SIMULINK calculation package [11].

The obtained results of calculations allowed to formulate practical conclusions for designers on how to adapt the machine to operate in elevated ambient temperatures.

References

1. Agnew, B., Sroka, Z.J.: A Study of the transient processes of a turbocharged diesel engine. In: Dudziński, P., Hawrylak, H., Hohl, G., Ketting, M. (eds) Off—Road Mechines and Vehicles in Theory and Practice. Proceedings of the International Conference Wrocław, Poland 23–24 Sept 1996. Technical University of Wrocław, Poland
2. Dudziński, P.: Theorie der Lenksysteme fuer industrielle Radfahrzeuge. Publishing House Oficyna Wydawnicza Politechniki Wroclawskiej, Wroclaw (2004)
3. Dudziński, P.: Lenksysteme fuer Nutzfahrzeuge. VDI – Buch. Springer (2008)
4. Dudziński, P., Chołodowski, J.: Energy efficiency of rubber tracked chassis. J. KONES **23**(2, 3), 97–104 (2016)
5. Girulski J.: Analiza wymiany ciepła w układzie chłodzenia zespołu napędowego pojazdu z silnikiem spalinowym (Heat Exchange in the Cooling System of the Power Unit with Combustion Engine) (M.A. Thesis) (in Polish). Politechnika Wrocławska, Wrocław (2002)
6. Grajnert J., Kwaśniowski S.: Biblioteka makroelementów SIMULINK-a do symulacji wymiany ciepła w pojazdach (Simulink's macroelements library for simulation of heat transfer in vehicles) (in Polish). IV Międzynarodowa konferencja naukowa w Polanicy Zdroju 1998, Materiały konferencyjne (Proceedings of the IV International Science Conference in Polanica Zdroj), Politechnika Wrocławska (1998)
7. Grajnert, J., Kwaśniowski, S., Sroka, Z. J.: Modelowanie dynamiki rozgrzewania silnika spalinowego podczas rozruchu (Modeling of the dynamics of the heating process of the combustion engine during starting) (in Polish). Materiały sympozjum na Politechnice Szczecińskiej (Proceedings of the Symposium at the Szczecin University of Technology), Szczecin, 23–24 June 1998
8. Kwaśniowski, S., Lewandowski, T., Słomski, W.: Zastosowanie metody bilansów elementarnych do komputerowej analizy obciążeń cieplnych zespołów napędowych (Application of the elementary balance method for computer analysis of thermal loads of drive units) (in Polish). III Konferencja naukowa w Kudowie Zdroju 11–14 grudnia 1996 (Proceedings of the III Science Conference in Kudowa Zdroj), Politechnika Wrocławska (1996)
9. Kwaśniowski, S., Sroka, Z., Zabłocki, W.: Modelowanie obciążeń cieplnych w elementach silników spalinowych (Modeling of thermal loads in elements of internal combustion engines) (in Polish). Publishing House Oficyna Wydawnicza Politechniki Wroclawskiej, Wroclaw (1999)

10. Ogrodzki, A.: Technika cieplna w pojazdach (Thermal technology in vehicles) (in Polish). Transport and Communication Publishers WKiŁ, Warszawa (1982)
11. SIMULINK. User's guide. Dynamic System Simulation Sowtware. The Math Works Inc
12. Skoczylas, A.: Przenoszenie ciepła (Heat Tranfer) (in Polish). Publishing House Oficyna Wydawnicza Politechniki Wroclawskiej, Wroclaw (1999)

Development Trends and Research Problems of Hydraulic Hammers for Mining and Civil Engineering

Marek Sokolski

Abstract The main areas of industrial applications of hydraulic hammers and basic development trends of these devices are presented. Basic functional structures and design structures of hydraulic hammers were characterized. On this background, the limit of further development of conventional hydraulic hammers was determined. Basic research problems have been formulated in relation to hammers used especially in mining and civil engineering. Examples of process analyzes in a hydraulic excitation system and in impact excitation system of conventional hydraulic hammers are presented.

Keywords Hydraulic hammers · Development · Topologic structures · Research problems · Case studies

1 Introduction

Hydraulic hammers are applied in many important industry branches such as the following:

- in mining and quarrying: rock mining without blasting, selective rock mining from side walls, secondary breaking oversized boulders, making underground galleries, quarrying rock blocks, etc.,
- in civil and building engineering: demolition of concrete foundations and buildings, removal of road surfaces, trenching in hard ground, etc.,
- in metallurgy, foundry and forging: hot slag removal, compaction of moulding sands, hammer forging, etc.

In many cases, especially in mining and civil engineering, hydraulic hammers are a reasonable alternative to techniques using explosives. Quite often, hydraulic hammers are the only rational solution that can be used, e.g., when subway tunneling or demolition works in dense urban areas. In such cases, the use of explosive materials

M. Sokolski (✉)
Faculty of Mechanical Engineering, Wroclaw University of Science and Technology, ul. Wybrzeze Wyspianskiego 27, 50-370 Wroclaw, Poland
e-mail: marek.sokolski@pwr.edu.pl

M. Sokolski (ed.), *Mining Machines and Earth-Moving Equipment*,
https://doi.org/10.1007/978-3-030-25478-0_9

can, in fact, cause damage to buildings or because of safety reasons, this could require the evacuation of people from areas where these works are being carried out.

In surface mining and quarrying, the serious hazards are caused by flyrock and lack from blasting operations which can cause injury or death to workers or other persons. In the most widely used technology of blasting by long vertical holes, the flyrock zone covers an area with a radius of $R = 200$ m, but in some cases, this range can reach up to $R = 500$ m.

Wide range of applications and varied operating conditions of hydraulic hammers create high demands on their durability and reliability. This applies in particular to the main working systems and their elements, i.e., to hydraulic excitation systems and impact excitation systems. Each repairing or replacing of elements of hammers in real operating conditions are usually quite difficult and expensive, and in some cases this is practically impossible to be carried out.

In this context, a need arises to formulate basic research problems, to identify the factors which have influence on the reliability of elements of impact excitation system as well as to develop a method of reliability assessment and prediction. However, modern hydraulic hammers are so complex objects that "intuitive" approach to these issues may not be sufficient.

In this chapter, some results of scientific and research author's activity are presented.

Because of adverse working conditions of hydraulic hammers, it is of special importance to improve their durability. It can be achieved through strengthening the surface. One of the newest methods which enables it, uses ion-plasma chromizing [1]. It was found out that in this way wear resistance can be increased by about 75% comparing with the surface without such machining. What makes such efforts more justified is the cost of tools of hydraulic hammers which can surpass 10% of the entire production costs [2].

2 Development of Hydraulic Hammers

The first industrial applications of hydraulic hammers were still in the 60 s of the twentieth century. The European pioneers in this field are Montabert (the first hydraulic hammer for concrete breaking in 1963) and Krupp (the first hydraulic hammer for rock mining in 1967).

The past 50 years were a period of dynamic development of hydraulic hammers, as a result of many years of experience and collaboration of designers, manufacturers, and users.

During this time, many sophisticated solutions have been created, such as control systems of amplitude of beater movement, "smart" systems of energy recuperation or unconventional hammers with hydraulic medium between the beater and the tool.

In particular, it can be noticed that the "hammer-tool" impact systems are a compromise solution to two mutually contradictory postulates [3, 4], i.e.:

- striving to maximize using the energy of the hydraulic medium, which is implemented by increasing the impact velocity of piston (beater),
- the need to reduce negative impact effects, i.e., excessive increase of stresses generated in the components of the "beater-tool" system, which requires limiting the impact speed.

The following companies (in alphabetical order): Atlas Copco AB (Sweden), Montabert S.A. (France), Krupp (Germany), Gullick Dobson (United Kingdom), Ingersoll Rand (USA), Joy (USA/Canada), Kone (Finland), Teledyne (Canada), Yamazen (Japan) were among the leaders who significantly contributed to the development of hydraulic hammers.

Most of these companies have developed series of hydraulic hammers with a versatile application, including hand hammers weighing 20–50 kg, as well as heavy hammers weighing 1000–1500 kg (or heavier) installed on stationary booms or as additional equipment of excavator or hydraulic loaders.

A detailed analysis of development trends of hydraulic hammers is presented by the author in work [4–6], where three following basic directions are defined:

- modernization strategies which involve the gradual improvement of the construction of hammers without changing the principle of operation of existing solutions. One of the typical examples of this procedure is to increase the technical "performance", especially the impact energy of hammers,
- diversification strategies which is characterized in the form of adaptation of hammers to specific, different operating conditions and user requirements. As a result of implementation of these trends, complete series of hydraulic hammers are being developed, which can optionally be used, for example to breaking of oversized boulders (by using the so-called moil point tools), removal of road surfaces (by using the spade tools), soil compacting (by using the compacting plate) etc.,
- innovative strategies, which are expressed by introducing qualitative modifications, which change partially or completely the principle of operation of existing solutions of hammers. As a result of this procedure, the new solution on a higher technical level is created—compared to conventional solutions—this offers new possibilities for the use of hammers.

The original concept of high energy hammers with a hydraulic fluid between the beater and the tool [7–9] is a significant example of those innovative trends. This generation of hammers is characterized by the ability to use much higher impact velocity. Therefore, impact energy in these hammers reaches a level up to 27 kJ, that is several times higher than the energy used in conventional hydraulic hammers of comparable sizes. This original solution the so-called HEFTI (High Energy Fluid Tappet Impactor), developed by Joy Manufacturing Co., is described in detail in [8, 9].

These trends are complementary and, de facto, they are different ways of carrying out the process of development of construction of hydraulic hammers. At the same time, it is characteristic that the introduction of only quantitative modifications, e.g., increasing the impact energy is effective, but only to a certain level, beyond which the effectiveness of the kind of development is lowered.

Usually, the ratio of the impact energy T_1 to the total mass of hammer m_H defined as follows:

$$W = \frac{T_1}{m_H} \tag{1}$$

is assumed as a basic comparative criterion of hammers.

Then, it can be concluded that increase of this index W for the next heavier and heavier hydraulic hammers is getting more difficult, or even impossible to obtain.

A detailed analysis of the development trends of hydraulic hammers was presented by the author in the work [4, 5].

3 Operational Structures of Hydraulic Hammers

Hydraulic hammers work in a specific environment as elements of a technical system with a specific structure and purpose.

The operational structure of a hydraulic hammer characterizes an energy conversion process that is occurring in the following, time- and space-arranged phases (Fig. 1):

- Phase I: conversion of the pressure (hydrostatic) energy of oil into the kinetic energy of piston (beater) and into the kinetic energy of oil stream in the working chamber of hammer,
- Phase II: conversion of the kinetic energy of beater and the hydrokinetic energy of oil stream in the working chamber into the impact energy of mass (mechanical) and into the energy of hydraulic impact,

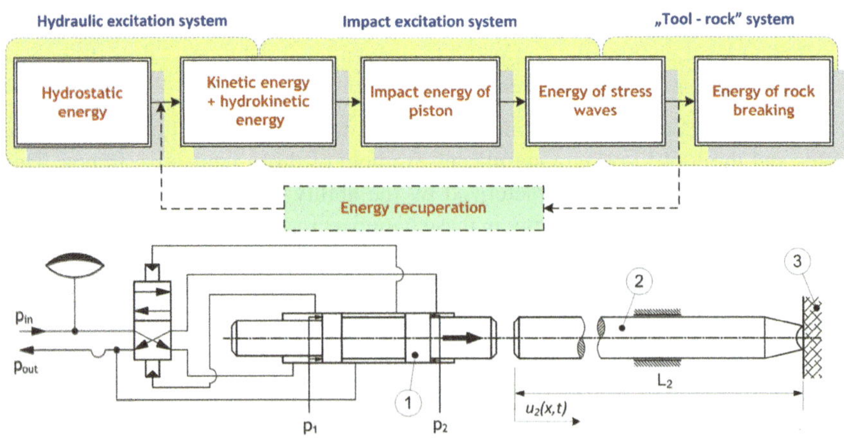

Fig. 1 Basic operational structure of a hydraulic hammer: 1—piston (beater), 2—tool, 3—target object (e.g., rock boulder or concrete foundation)

- Phase III: conversion of impact energy into the energy of stress waves generated in the piston (beater) and the tool,
- Phase IV: conversion of the energy of elastic stress waves generated in the tool into the useful energy of rock breaking (destruction of the rock structure).

Some modern hydraulic hammers have special systems to recover some portion of the energy not used during the working stroke. Then, in the operational structure of hammer, an additional stage of the so-called energy recuperation during the return stroke appears. Such solutions are used, e.g., in some Krupp hydraulic hammers.

Among the most crucial parts of hydraulic hammers are reversing valves and accumulators as both of them control the operation of these devices. Example of an innovative steering method is presented in [10]. The main benefit of its utilization is a considerable decrease of reaction time. In [11] the research shows the most advantageous value of so-called precharge pressure compared to working pressure, which should be up to 80%. It is worth noting that this proportion can be controlled by selected parameters of accumulators.

4 Constructional Structures of Hydraulic Hammers

There are the following subsystems in the design structure of a modern hydraulic hammer:

- hydraulic excitation system, the main purpose of which is to convert the energy supplying the hammer to the kinetic energy of the striking element (piston) and to the hydrokinetic energy of the stream of liquid in the working chamber of the hammer. This excitation system includes an internal control subsystem and is additionally equipped with a hydraulic accumulator for the supply of power stroke. This accumulator is usually located in the hammerhead (some hammers have an additional accumulator to relieve the pressure pulsation in the return line),
- impact excitation system in which the kinetic energy is transformed into the energy of waves generated in the tool; conventional hydraulic hammers have hammer-type impact excitation systems,
- main cylinder inside of which there are hydraulic excitation system and impact excitation system located,
- shock absorber and vibration damping system located in the casing of hammer. In order to reduce the dynamic reaction of impact excitation system on the supporting boom, the hammer's working cylinder is not rigidly integrated with the housing, but is embedded on internal vibro-isolators (i.e., on systems of elastomeric springs and cushions). Thus, the energy transferred to the supporting boom does not usually exceed 10–15% of the total impact energy; this fact was confirmed by the extensive research in industrial conditions [8, 9] (also by author's research in Polish copper mines).

- hammer housing in which the working cylinder is slidably mounted. In the housing, there are connection sockets for external hydraulic conduits (supply and return) and connection holes for mounting the hammer to the support boom or manipulator,
- support system: boom or manipulator made in stationary version or on a mobile chassis (wheeled, crawler or rail).

4.1 Basic Structures of Hydraulic Excitation Systems

The basic task of the hydraulic excitation system is to generate a cyclic movement of the piston (beater), with appropriate energy and frequency. In this system, the hydrostatic energy is converted to kinetic energy of the piston and to kinetic energy of liquid flow in the working chamber. The main elements of the hydraulic excitation system—which is one of the major subsystems of hammer—are internal flow dividing valve, working chambers, bladder type accumulator (or two accumulators) and internal connecting channels.

The basic elements of the hydraulic excitation system are the distribution valve subassembly, working chambers, accumulator (or two accumulators) and internal connecting channels.

One of the main characteristic features of the hydraulic excitation systems, regardless of design variants of hammers, is the fact that the working stroke is always supported by hydraulic accumulator.

The following three basic topological structures of the hydraulic excitation systems can be perceived [4, 5]:

- excitation systems with continuously supplied back working chamber by oil flow under high-pressure (Fig. 2a). The front working chamber is alternately attached to the return line (working stroke) or supply line (return stroke),
- excitation systems with the continuously supplied front working chamber by oil flow under high pressure (Fig. 2b). The back working chamber is alternately attached to the supply line (working stroke) or return line (return stroke),
- excitation systems with alternately supplied working chambers by oil flow under high pressure or connected with the return line (Fig. 2c). During the working stroke, the back chamber is supplied by high pressure and the front chamber is connected to the return line. During the return stroke, it is in inverse: the front chamber is supplying by high pressure and the back chamber is attached to the return line.

In practice, the first two of the previously mentioned systems (Fig. 2a and b) are usually used. In particular, the hydraulic excitation system with continuously supplying of the back chamber is typically used in medium size hammers (Fig. 3).

In contrast, the hydraulic excitation systems with continuously supplying of the front chamber are most commonly used in heavy hammers (Fig. 4). One of the main advantages of this structure is better shock damping in the front dead centre of beater (piston).

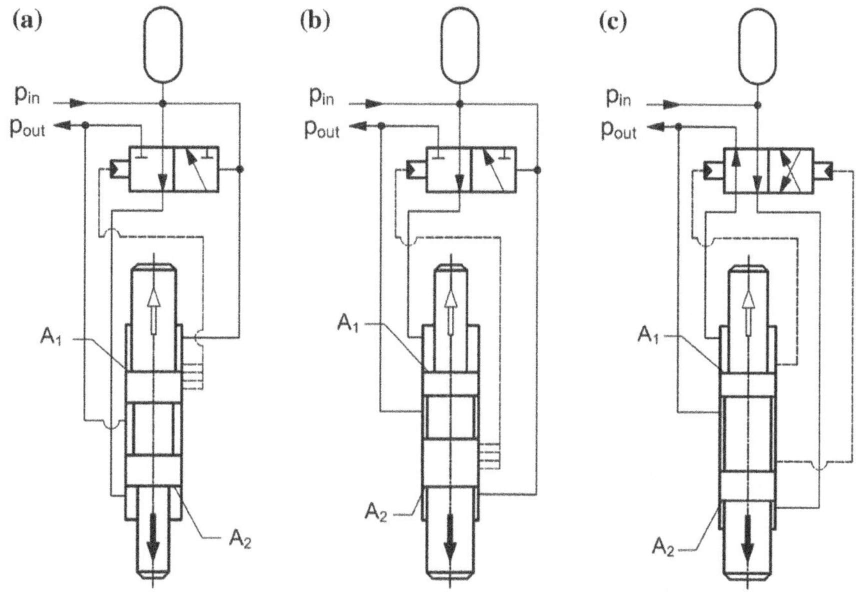

Fig. 2 Basic topological structures of hydraulic excitation systems of hammers: **a** continuously supplying of the back working chamber, **b** continuously supplying of the front working chamber, **c** alternately supplying of the working chambers, A_1, A_2—active areas of working chambers

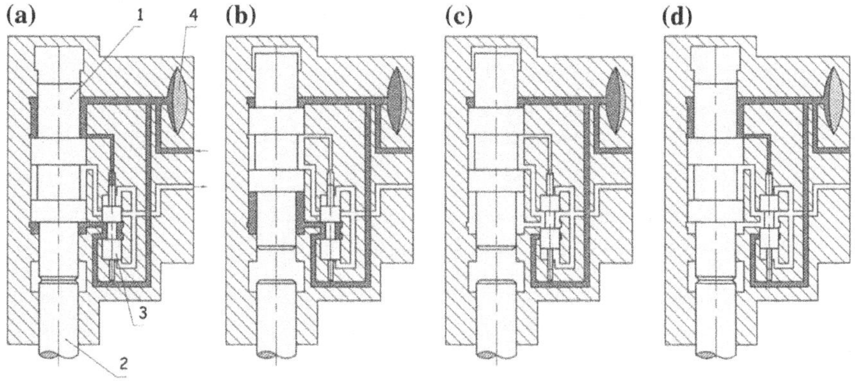

Fig. 3 Hydraulic excitation system with continuously supplying of the back working chamber: **a**, **b** start and end of return stroke, **c**, **d** start and end of working stroke; 1—piston (beater), 2—tool, 3—distribution valve, 4—accumulator (author's own work based on [5] and materials from "Krupp")

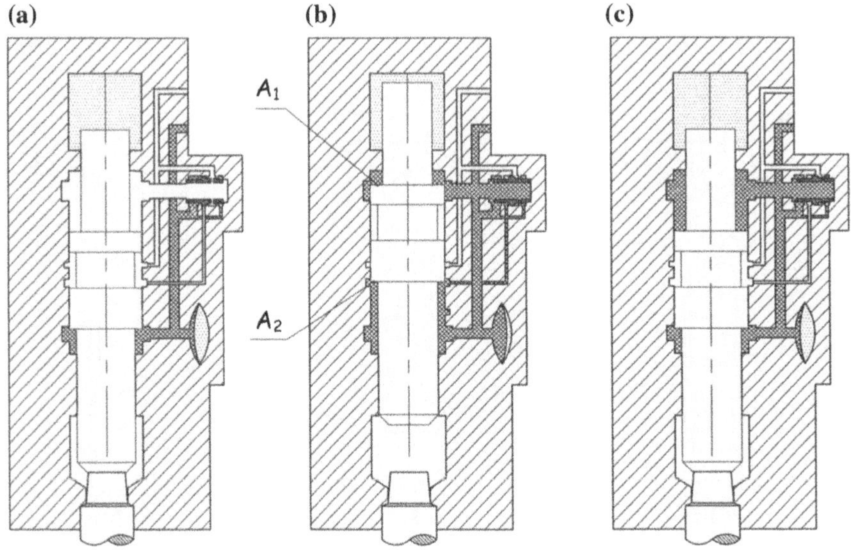

Fig. 4 Hydraulic excitation system with continuously supplying of the front working chamber: **a** start of return stroke, **b**, **c** start and end of working stroke; 1—piston (beater), 2—tool, 3—distribution valve, 4—accumulator (author's own work based on [5] and materials from "Krupp")

As a part of the diversification activities, consisting in striving for better adaptation of hydraulic hammers to different operating conditions, there are also additional amplitude control systems of the piston stroke (on two or three levels) applied. Then, the impact energy of hammer is changed proportionally to the value of piston stroke.

The use of such a system enables better flexible adjustment of the parameters of the impact process to the operating conditions by:

- increasing the frequency of impacts while reducing energy,
- reducing the frequency of impacts while increasing energy.

Thus, for example, a hammer that breaks up rock lumps at the grate screen station can work with less energy of impact (the so-called "short stroke"), and in the case of rock mining from side walls the hammer works with maximum energy (the so-called "long stroke" of the beater). Examples of such solutions are presented by the author in the works [4, 5].

4.2 Basic Structures of Impact Excitation Systems

In contemporary hydraulic hammers, the different systems of impact excitation are used. According to the classification proposed by the author, the following, basic, topological structures can be perceived (Fig. 5) [4, 5, 8, 9]:

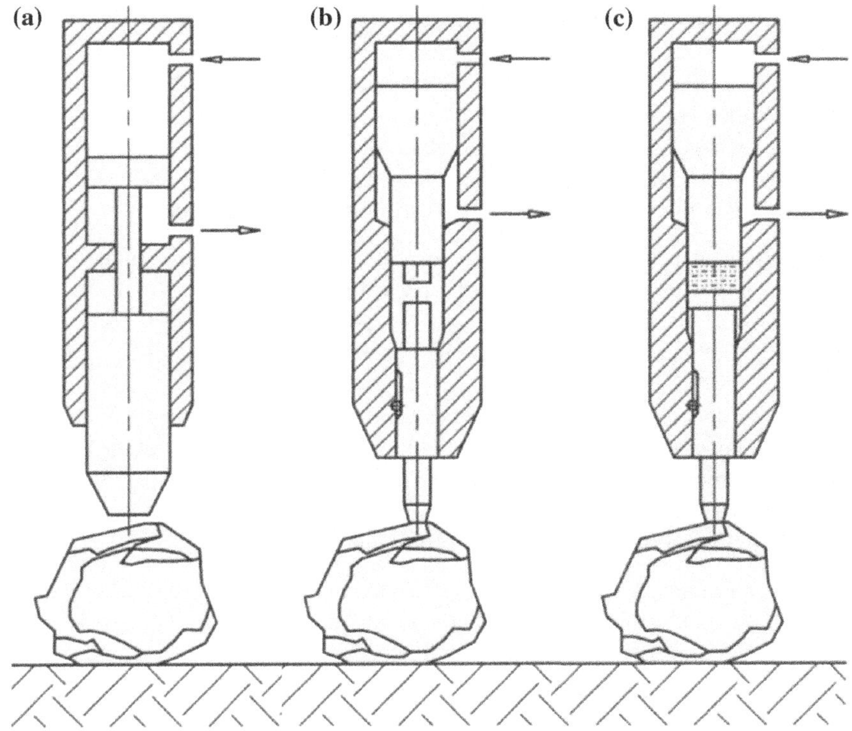

Fig. 5 Basic structures of impact excitation system of hammers: **a** projectile type structure, **b** hammer-type structure with direct impact, **c** hammer-type structure with indirect impact (author's own work based on [5, 8, 9]

- Projectile type excitation systems (Fig. 5a), where the impact system is made as one piece system which also acts as the beater and as the tool. In such hammers, the working process is carried out by the tools which impact directly against the rock.
- Hammer-type excitation systems in which the functions of beater and tool are separated, herein, depending on the nature of the impact process, the following structures can be perceived:
  - Hammer-type excitation systems with direct impact (Fig. 5b) in which the rear faces of the tool are directly struck by a beater (there is a so-called "metal–metal" impact),
  - Hammer-type excitation systems with indirect impact (Fig. 5c) in which there is applied an additional medium between the beater and the tool (the beater does not strike directly against the tool). Such an original solution with a hydraulic medium, i.e., a fluid tappet, has been applied in the HEFTI hammers developed by Joy [8, 9].

4.2.1 Projectile-Type Structure of Impact Excitation Systems

One of the basic advantages of impact excitation systems with a projectile type structure is the possibility of using high impact energy T_1. The permissible value of this energy, which is limited by the strength of tool material, can be estimated on the basis of the interaction analysis between the tool and the rock in the following way.

According to this idea, the following basic assumptions are made:

- the relationship between the resistance force F and the deformation h of the contact surface "tool–rock" has the form of the linear function:

$$F(h) \cong k_h \cdot h \tag{2}$$

(where: k_h denotes coefficient of stiffness of the "tool–rock" contact).

- the impact energy is equal to the energy of tool deformation of the at the moment when maximum stress σ_{max} occurs (energy dissipation in the impact system is disregarded),
- the tool has a rod configuration with a constant cross-section A_2.

For these assumptions, the following estimation of the permissible value of impact energy $T_{1(max)}$ in a system with a projectile type structure is obtained [4, 5]:

$$T_{1(\max)} = \frac{1}{2} \cdot \frac{\sigma_{perm}^2 \cdot A_2^2}{k_h} \tag{3}$$

As an application example, the following case (developed by the author) of assessment of the maximum permissible impact energy $T_{1(max)}$ is presented. It was assumed that [4, 5]:

- working tool has a diameter $d = 100$ mm and a wedge-shaped tip with angle $\alpha = 60°$,
- permissible stress for the tool material (taking into account the required safety factor) is $\sigma_{perm} = 200$ MPa,
- rock to be broken is a granite one; the coefficient stiffness of the "wedge-rock" contact is $k_h = 80$ MN/m.

It results from the formula (3) that the upper limit of impact energy that can be used in this system is equal to $T_{1(max)} = 15.5$ kJ.

However, one of the significant disadvantages of the projectile type structure of impact systems is the fact that the tool is not pre-pressed against the rock when mining. As a consequence, there is a recoil phenomenon which is much stronger than in impact system with hammer-type structure.

This has a negative influence on both the efficiency of the mining process as well as on the dynamic loads of supporting boom. Especially, when hard rock mining, the recoil of the tool can absorb a significant portion of the impact energy. This causes high additional dynamic forces acting on a boom and thus reducing the value of the energy which can be used for rock mining [4, 5].

In general, for these reasons, hammers with projectile type structures of impact excitation systems do not use in mining. However, they are applied in civil engineering, e.g., for removing road surfaces or trenching in hard ground.

4.2.2 Hammer–Type Structure of Impact Excitation

Hammer-type structures with direct impact are the most commonly used structures of impact excitation systems of hydraulic hammers. The essence of the operation of these systems is the impact phenomenon generated cyclically in the "piston-tool" working system, without additional intermediary elements between piston and tool.

Striking systems of this structure have many advantages, among others, the possibility of applying a preliminary pressure of the tool against the target (rock, concrete, etc.) and then the unfavorable phenomenon of recoil can be limited.

However, from the perspective of the past half-century, the specific symptoms of the "aging" of this type generation of hammers appear. This fact has been noticed by the author and described in details in the works [4, 5].

This is primarily about the barriers that are more clearly perceived to further develop this generation of impact excitation systems and they are resulting primarily from the need to:

- limit the impact velocity,
- limit the piston (beater) mass.

Permissible impact velocity v_1 in hammer-type structure with direct impact systems is quite strongly limited by the material strength of the piston and tool.

The explanation of this issue is provided by the classic, one-dimensional wave theory of impact. Based on this theory, in the case when two Hooke's bodies with bar configuration with constant cross-sections and flat face are colliding; the following relationships can be formulated [4, 5, 8]:

$$\sigma_1 = \frac{E}{1 + \frac{A_1}{A_2}} \cdot \frac{v_1}{c} \quad ; \quad \sigma_2 = \frac{E}{1 + \frac{A_2}{A_1}} \cdot \frac{v_1}{c} \tag{4}$$

where: σ_1 i σ_2—final stresses in the beater and the tool, respectively, v_1–impact velocity, A_1, A_2—cross-sectional area of piston (beater) and tool, respectively, c, E—speed of sound and Young's modulus of steel, respectively,

Most often, the beater and tool have the same cross-sections (i.e., $A_1 = A_2$). In addition, for typical steel, Young's modulus is equal to $\mathrm{E} = 2.1 \times 10^5$ MPa and the wave speed is $c = 5150$ m/s.

In such conditions, the impact velocity $v_1 = 10$–12 m/s generates in the piston and tool the stresses which are equal to $\sigma_1 = \sigma_2 = 200$–$250$ MPa.

According to the author's knowledge, these values of normal stresses are usually accepted as a permissible level. This is a conscious simplification of the issue, among others because the destruction of material under impact load is also caused by some additional phenomena, such as material fatigue, low-frequency cracking, local plasticity, local stress concentration, and others that they are usually omitted.

As a consequence of these reasons, there is a need to limit the impact velocity in conventional hammer to values of $v_1 = 10\text{–}12$ m/s.

In turn, the piston mass m_1 is limited by the acceptable level of recoil energy which is reflected from the rock to be broken. In fact, if we disregard the influence of the force of the initial pressure of the tool against the rock, then the recoil energy T_{rec} can be estimated with sufficient accuracy by the following expression:

$$E_{rec} = \frac{m_1}{m_H} \cdot T_1 \tag{5}$$

where: m_H, m_1—denote hammer mass and piston (beater) mass, respectively.

Comprehensive tests of hydraulic hammers in industrial conditions [8, 9] as well as the author's research the Roxon 602 hammers carried out in the Polish copper mines [5] have shown that especially due to the durability of the steel structure of the booms, the recoil energy T_{rec} should not exceed 10–15% of the total impact energy T_1. Therefore, as it follows from the formula (5), the same order should be the piston mass m_1 in the total hammer mass of the hammer m_H.

On the other hand, the postulate to ensure sufficient maneuverability of the hammer imposes a limitation of its total mass m_H. It is especially important in the case of mobile hammers, installed as additional equipment on the booms of excavators or front loaders. Although this issue has not yet been clearly settled, there are some indications that the maximum total mass of the hammer should not exceed the value of $m_H = 1000\text{–}1500$ kg [4, 5].

Summing up, if both above-mentioned limitations of impact velocity v_1 and piston mass m_1 are associated, it can be found that the rational development barrier of conventional hydraulic hammers is determined by the impact energy of $T_{1(max)} =$ 5.7 kJ. This conclusion formulated once by the author [4] still remains valid.

4.2.3 Hammer–Type Structure with no Steel on Steel Impact Excitation Systems

The limitation of the impact velocity to the level of $v_1 = 10\text{–}12$ m/s, occurring in conventional hydraulic hammers, was one of the basic premises to develop the concept of new-generation hammers.

The idea of using a hydraulic medium that separates both metallic elements of the impact system piston and tool turned out to be really a breakthrough solution. This solution was used in hydraulic hammers named HEFTI (Joy Mfg. Co.) [7–9], which have the energy index $W = T_1/m_H = 8.5\text{–}25.5$ J/kg, and in this respect, the hammers of this new generation "totally beat" conventional hammers.

The basic advantage of eliminating the "metal on metal" impact and replacing it by a hydraulic medium system is the possibility of using much higher impact velocity v_1 with a relatively lower level of stress generated in the impact system.

An extensive analysis of stresses in the impact system "beater—hydraulic medium—tool" is presented in [8, 9] and also in the works of the author [5].

Another advantageous feature of this type of impact systems is increasing the durability of components of the working system because the contact stresses are more uniformly distributed and local stress concentration is reduced.

Further advantages of the new generation of hydraulic hammers are the reduced activity of vibroacoustic processes in the impact zone, in particular, the overall noise level of hammer during operation is relatively low and the recoil energy is also reduced to a value not exceeding several percent of impact energy [4, 5, 8, 9].

However, in the general assessment of this new generation of hydraulic hammers it should be taken into account that there could arise some problems concerning the seals working under high-pressure conditions of p = 100–200 MPa. A detailed description of this generation of hydraulic hammers was presented by the author in [4, 5].

5 Basic Research Problems of Hydraulic Hammers

In the aspect of complex identification of the working process of hydraulic hammers and diagnosing the technical condition of these devices, the following main research problems are to be solved [4, 5]:

- analysis of the process of transformation of hydrostatic energy into the kinetic energy of piston (beater) and hydrokinetic energy of liquid stream in the working cylinder of hammer taking into account the main parameters of the hydraulic excitation system,
- analysis of the impact process carried out in the "piston-tool" system, taking into account the basic parameters of the impact excitation system,
- analysis of the dynamic interaction between the hammer and supporting boom in the context of the assessment of fatigue strength of boom construction under impact loads,
- analysis of residual processes in the context of assessment vibroacoustic hazards as well as diagnosis of the technical condition of working systems of hammers,
- analysis of the target process (rock mining, concrete breaking, etc.) under impact loads in the context of proper selection of tools.

The analysis and finding proper solutions of the above-mentioned research problems enable the formulation of basic characteristics of the working process of hydraulic hammers and as well as basic characteristics of residual processes.

Wherein, the following characteristics of hydraulic hammers are particularly important:

- the kinematic characteristic $\boldsymbol{\Phi}_{\mathrm{k}}$ expressing the final velocity v_1 of hydraulic impact excitation as a function of basic controllable input quantities $\boldsymbol{X}$ and non-controllable parameters $\boldsymbol{P}$ of hammer,

- the energy characteristic $\mathbf{\Phi}_E$ expressing the final kinetic energy T_1 of hydraulic impact excitation as a function of basic controllable input quantities **X** and non-controllable parameters **P** of hammer,
- the force characteristic $\mathbf{\Phi}_F$ expressing the maximum impact force F_{max} as a function of basic controllable input quantities **X** and non-controllable parameters **P** of hammer,
- impulse characteristic $\mathbf{\Phi}_S$ expressing the impulse S_1 of impact force as a function of basic controllable input quantities **X** and non-controllable parameters **P** of hammer,
- the stress characteristic $\mathbf{\Phi}_\sigma$ expressing the state of stress σ_{eq} in impact excitation system as a function of basic controllable input quantities **X** and non-controllable parameters **P** of hammer.

The above-formulated characteristics of hydraulic hammer can be written in a general form as:

$$v_1 = \Phi_k(X, P); T_1 = \Phi_E(X, P); F_{max} = \Phi_F(X, P)$$
$$S_1 = \Phi_S(X, P); \sigma_{eq} = \Phi_\sigma(X, P) \tag{6}$$

An opinion has been established for a long time that the impact energy T_1 is a main quantity of hydraulic hammers.

This is justified by the fact that in many cases the conditions for breaking the rock or concrete structure are formulated in terms of energy. It is enough to mention, for example, the classic Griffith theory of material fracture or the rock crushing theories using Kick and Rittinger's laws.

Therefore, the impact energy T_1 is almost always given in the catalogs of hammers and in practice this parameter is most often the basis for hammer selection to operating conditions.

According to the author, the experimental assessment of impact energy of hammer is one of the most difficult tasks.

One of the basic reasons for this was the fact that until recently there were no uniform standards for determining the useful energy of hammers. The applied research methods were shrouded as a specific "company secret". Therefore, often the results of impact energy measurements of the same hammers could be differed in values, depending on the conditions and used research method.

It was not until the second half of the 1990s that the Mounted Breaker Manufacturers Bureau (MBMB) in the USA was created, which dealt with the development of rules for defining the basic operational parameters of hydraulic hammers.

The so-called CIMA/MBMB method has been developed for determining impact energy, based on tool deformation values ε, measured at the recommended distance $x = 1.2 \cdot L_1$ from the tool tip (L_1—piston length). The measurement should be performed by using strain gauges connected in Wheatstone full bridge system (or half-bridge system).

An original method of determining the impact energy of hammers is presented in the work [12]. The idea is based on measurements of impact energy by means of force converter with an embedded elastic element. This method has been applied in Polish copper mining industry [13].

A slightly different approach to the effectiveness of rock breaking process is presented by theories that assume that the necessary condition for material decohesion is to exceed the so-called local strength threshold. This requires achieving a sufficiently large impact force F_{max} in the hammer.

In turn, the advisability of considering the impact force impulse S_1 as a factor influencing on rock breaking process is justified by the following example.

The rifle bullet 7.62 mm with mass $m_1 = 11.8$ g, with an initial velocity of approx. $v_1 = 780$ m/s, reaches kinetic energy of $T_1 = 3600$ J. This value is more than twice as high, for example as the energy $T_1 = 1600$ J of the Roxon 602 hydraulic hammer. Nevertheless, the effect of rock breaking with a rifle bullet is practically negligible. The question then arises, why is this happening?

This apparent paradox can be explained if we assume an additional hypothesis that the effectiveness of impact mining depends not only on the value of total energy T_1, but also on the value of impact force impulse S_1 of bullet or tool.

The relationship between these quantities results from Kelvin's theorem about the work of impulsive forces and has the following form [4, 5]:

$$S_1 = \sqrt{2T_1 \cdot m_1} \tag{7}$$

where: m_1—denotes piston (beater) mass or bullet mass

The small efficiency of rock breaking by using a rifle bullet can be justified by the fact that despite a significant kinetic impact energy $T_1 = 3600$ J, the bullet has a very small mass $m_1 = 11.8$ g. As a consequence, as it arises from formula (7), the bullet has also a small momentum value $m_1 \cdot v_1 = 9$ Ns which is finally converted to small value of force impulse $S_1 = 9$ Ns. This value is insufficient to overcome the local rock strength threshold.

In contrast to this, the Roxon 602 hammer has a piston with a mass of $m_1 = 30$ kg, which at the moment of impact gives a force impulse of $S_1 = 310$ Ns, so this value is many times greater than this one in the case of a rifle bullet.

The above example clearly proves the purposefulness of considering the force impulse S_1 as an important factor of hydraulic hammer.

The formulation of the basic characteristics of hydraulic hammers should be one of the results of solving these above-mentioned research problems, whereby two basic methods are possible:

- analytical or numerical studies using appropriate physical models or numerical models of working systems of hammers,
- experimental studies using research results obtained in the laboratory and/or in real operating conditions.

It is obvious that each of these methods has its advantages and disadvantages as well as its supporters and skeptics.

One of the basic items in the subject literature on analytical modeling of hydraulic drive systems of presses and hammer is the work [14]. Despite the passage of time, many of the results and conclusions contained therein are still valid.

In turn, the examples of an analytical and experimental approach to modeling the working process of hydraulic hammers are works [15, 16].

The examples of a "new wave" in the modeling of hydraulic hammers are the works [17, 18] where some computer simulation packages were used.

The subject matter of hydraulic hammers for mining and civil engineering is present at the Faculty of Mechanical Engineering of the Wroclaw University of Science and Technology from the beginning of the 70s of the twentieth century. Within this activity, analytical studies and research of hammers in the laboratory as well as in real mining conditions are carried out.

In this chapter two examples of this activity are presented. The first example is related to the theoretical analysis of hydraulic excitation systems and the second example concerns the numerical simulation of general state of stress in impact excitation systems.

6 Analysis of the Hydraulic Excitation System

In this chapter, as an example #1, the analytical study of hydraulic excitation system is presented. The basic physical model (Fig. 6) meets the following basic assumptions:

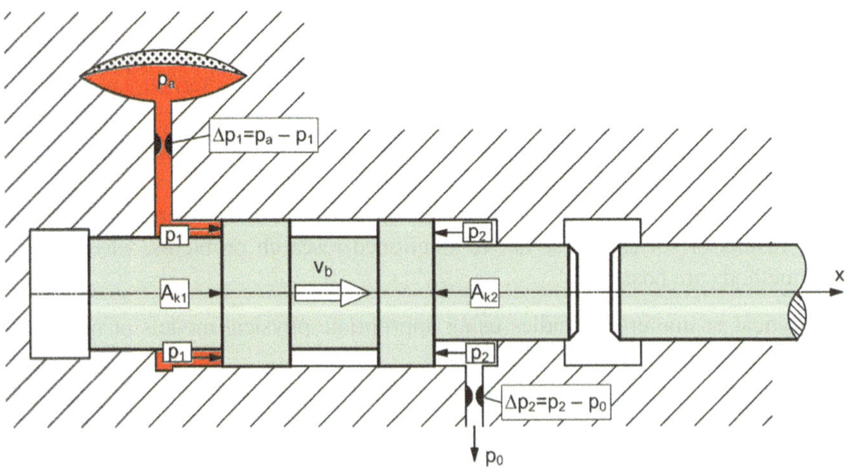

Fig. 6 Physical model of hydraulic excitation system of hammer: p_1, p_2—pressures in working chambers, A_{k1}, A_{k2}—active areas of working chambers

- the excitation system has the structure with continuous supply rear working chamber, i.e.:
  - the rear working chamber is still connected to the supply line and is supplying by high-pressure oil stream,
  - the front working chamber is alternately connected to the return line (working stroke) or to the high-pressure line (return stroke),
- in the supply and return lines, the continuity conditions of the liquid stream are performed (no leakage at leaks),
- the oil flows in the supply and return lines are turbulent, i.e., the pressure losses are proportional to the second power of the flow velocity,
- the working stroke is supported from a bladder type accumulator installed in the supply line,
- the process of accumulator discharging during the working stroke is polytropic with a constant exponent κ.

The other assumptions were described in detail in the author's work [5].

The movement of piston during the working stroke in the physical model of the hydraulic excitation system (Fig. 6) is described by the following equation:

$$m_r \cdot \frac{\mathrm{d}v_b(t)}{\mathrm{d}t} = p_1(t) \cdot A_{k1} - p_2(t) \cdot A_{k2} - R_T \tag{8}$$

where: v_b—temporary piston velocity, m_r—total mass of piston and fluid flow in working chambers of hammer, R_T—total mechanical resistance of piston motion (frictional resistance on cylinder bearing surface and seals).

6.1 Analysis of the Hydraulic Excitation System with Constant Supply Pressure

If in Eq. (8) an additional assumption is made that during the working stroke the pressure p_1, p_2 in the working chambers are independent of time, i.e., $p_1 = \text{const}$ and $p_2 = \text{const}$, then the piston motion is described by the following Riccati differential equation with constant coefficients:

$$\frac{\mathrm{d}v_b(t)}{\mathrm{d}t} = \frac{F_{r(0)}}{m_r} - \frac{R_\xi}{m_r} \cdot v_b^2(t) \tag{9}$$

where: $F_{r(0)}$ denotes the nominal active force of working stroke, R_ξ is total hydraulic resistance per second power of piston velocity.

This active force $F_{r(0)}$ in working stroke of a piston is determined as follows:

$$F_{r(0)} = p_{a(0)} \cdot A_{k1} - p_0 \cdot A_{k2} - R_T \tag{10}$$

where: $p_{a(0)}$ denotes the pressure delivered from accumulator (it is assumed that $p_{a(0)}$ = const).

Then, the mathematical model (9) of a piston motion is done by the differential equation with separated variables. The solution of the motion Eq. (9), for zero initial conditions, is conveniently to present in the form of the following dimensionless formulas [5, 14]:

$$\left.\begin{array}{c} s_{bj}(t) = \frac{s_b(t)}{\Phi_s} = \ln[\cos h(\tau)]; \quad v_{bj}(t) = \frac{v_b(t)}{\Phi_v} tgh(\tau) \\ a_{bj}(t) = \frac{a_b(t)}{\Phi_a} \frac{1}{\cos h^2(\tau)} \end{array}\right\} \tag{11}$$

where: s_b, v_b, a_b denotes displacement, velocity, and acceleration of piston, respectively. The symbol τ denotes relative time, defined as $\tau = t/\Phi_t$.

The symbols Φ_s, Φ_v, Φ_a in the solutions (11) denote displacement scale, velocity scale, and acceleration scale, respectively, and are defined as follows:

$$\left.\begin{array}{c} \Phi_t = \frac{m_r}{\sqrt{R_\xi \cdot F_{r(0)}}} \quad ; \quad \Phi_s = \frac{m_r}{R_\xi} \quad ; \\ \Phi_v = \frac{\Phi_s}{\Phi_t} = \sqrt{\frac{F_{r(0)}}{R_\xi}} \quad ; \quad \Phi_a = \frac{\Phi_s}{\Phi_t^2} = \frac{F_{r(0)}}{m_r} \end{array}\right\} \tag{12}$$

These dimensionless solutions are presented in Fig. 7.

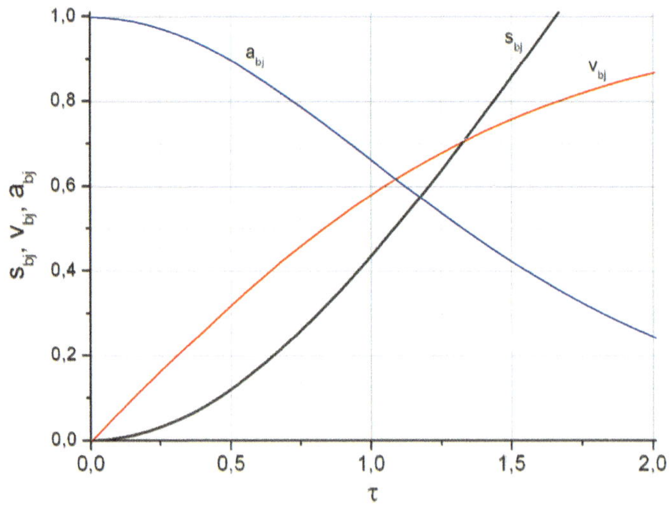

Fig. 7 Dimensionless characteristics of piston motion in working stroke under constant pressure

The classic dimensionless characteristics (Fig. 7) give an overall qualitative view of changes in the basic kinematic quantities of the hydraulic excitation process, i.e., in particular on the velocity and acceleration of piston. It results from these characteristics, among others, that under the constant working pressure in the working chamber (generated by an accumulator) the unit velocity of piston v_{bj}, expressed in relative values, increases from the initial value $v_{bj}(0) = 0$ and aims asymptotically to the value $v_{bj} = 1$. The increase of the piston velocity is not constant and it is proceeding with a decreasing acceleration. The most intensive increase of velocity is occurring in the initial phase of motion, i.e., when the relative time is $\tau \leq 1.0$, then the relative acceleration is $a_{bj} \geq 0.4$.

It can be seen that during the working stroke the piston velocity v_b goes asymptotically to the following limit value:

$$v_b(t) \rightarrow v_{b(max)} = \varphi_v = \sqrt{\frac{F_{r(0)}}{R_\xi}} \tag{13}$$

Hence the important conclusion is that the final impact energy $T_{b(max)}$ of piston, under the constant active pressure in the working chamber is:

$$T_1 = T_{b(max)} = \frac{1}{2} m_b v_{b(max)}^2 = \frac{1}{2} m_b \frac{F_{r(0)}}{R_\xi} \tag{14}$$

6.2 Analysis of the Hydraulic Excitation System with Polytropic Supply Pressure

The assumption that during the working stroke the active pressure generated by an accumulator is constant (i.e., $p_a = \text{const}$) can be justified only for cases when the accumulator capacity is much larger than the swept volume V_{swe} of the working cylinder, i.e., when $V_a \gg V_{swe}$. However, this condition is not always met.

Therefore, in further analysis, it was assumed that during the working stroke, the pressure generated by accumulator is not constant, but it results from the polytropic expansion of accumulator and the polytropic exponent is equal to $\kappa = 1.4$.

Under such conditions of hydraulic excitation process, the mathematical model of piston motion can be expressed by the following differential equation [5]:

$$\frac{d\dot{s}_b(t)}{dt} = \frac{F_{r(0)}}{m_r} \frac{1}{\left[1 + \frac{1}{\sqrt[\kappa]{k_{p0}}} \cdot \frac{1}{k_{Va}} \cdot \frac{s_b(t)}{s_{b0}}\right]^\kappa} - \frac{R_\xi}{m_r} \cdot \dot{s}_b^2(t) \tag{15}$$

where

s_{b0} value of free displacement of piston

p_{G0} precharge accumulator pressure
k_{p0} coefficient of precharge pressure
k_{Va} coefficient of accumulator volume and:

$$k_{p0} = p_{G0}/p_a \; ; \; k_{Va} = V_a/V_{swe} \tag{16}$$

The dimensionless solutions of the Eq. (15) which determine the displacement, velocity, and acceleration of piston under polytropic expanding working pressure are presented in Fig. 8.

A comparison of the calculation results for the variant of constant supply pressure and the variant of polytropic variable pressure is shown in Fig. 9.

The comparison between the analysis results obtained from both models of hydraulic excitation gives, inter alia, the following conclusions:

- the assumption that the supply pressure during the working stroke is constant is justified only for a sufficiently large battery capacity, practically when the condition $k_{Va} = V_a/V_{swe} \geq 5.0$ is met (Fig. 9),
- for conditions where $k_{Va} < 5.0$ the differences between the results obtained on the basis of both models (9) and (15) are significant.

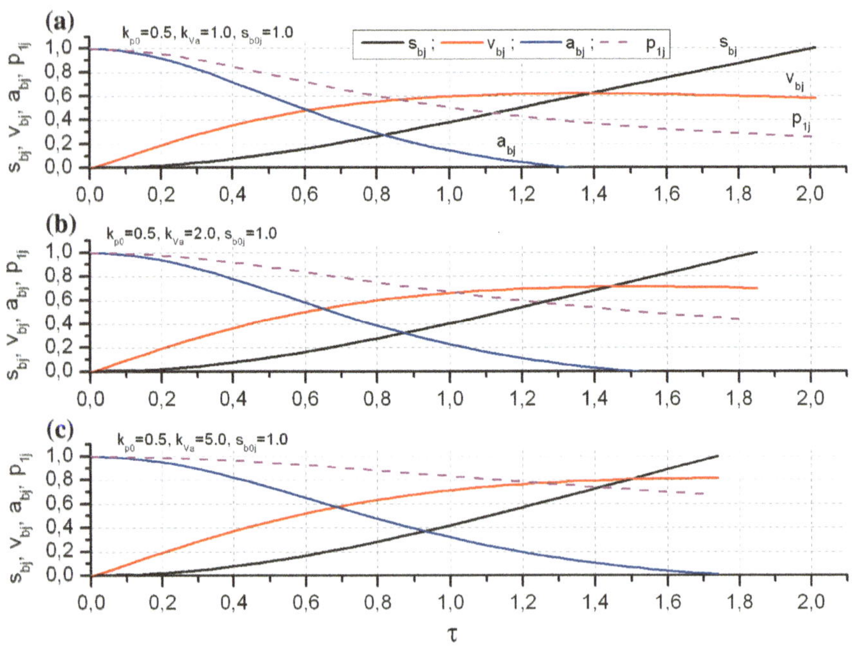

Fig. 8 Dimensionless characteristics of piston motion in working stroke under polytropic expanding pressure

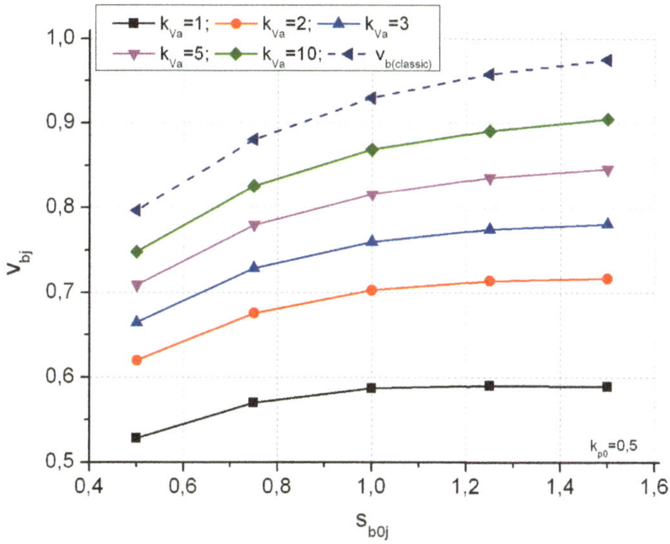

Fig. 9 Unit velocity of piston v_{bj} calculated on the basis of constant pressure model (Eq. (9), dashed line) and on the basis of the model of polytropic expanding pressure (author's model, continuous lines) for several values of coefficient of accumulator volume k_{Va}

7 Analysis of the Impact Excitation System

As a second example, some results of the numerical study of impact excitation process in a "piston-tool" system are presented. The developed physical and numerical models (one of them is shown in Fig. 10) refer to the real working systems of the Roxon 602 hydraulic hammers, which were tested by the author in the conditions of the Polish copper mines.

The following main goals of numerical simulations were assumed:

- the strength analysis of the impact system with especially focus on the so-called critical zones of tool,
- the estimation of the maximum permissible impact velocity which does not cause the damage of elements of impact system,
- the determination of the factor of safety (FoS) for a given impact velocity.

The equivalent von Mises stress σ_{red} was used as a measure of local state of stress caused in elements of impact system.

The first boundary condition BC1 ($z = L_1$, Fig. 10) which determines the interaction between the beater and the tool is based on the model of surface to surface hard contact in normal direction without friction.

The second boundary condition BC2 ($z = L_1 + L_2$, Fig. 10) which determines the interaction between the tool and the rock is based on an extreme case when the single impact does not break the rock.

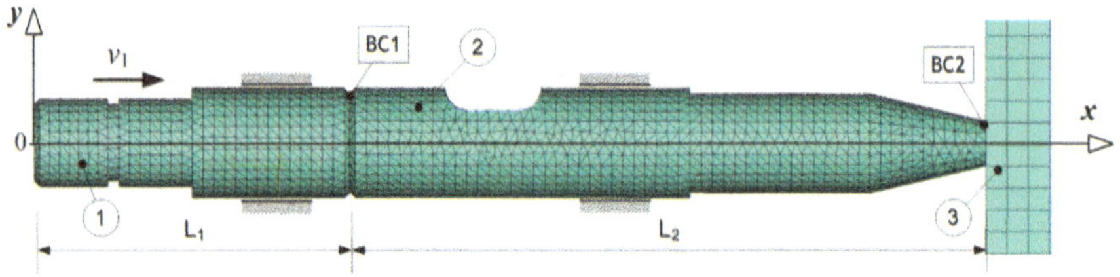

Fig. 10 Model of impact system for the FEM simulations: 1—piston, 2—tool, 3—rock, BC1, BC2—boundary conditions

Some results of the FEM simulation of the normal stress σ_Z on the surface of the impact zone, for three values of impact velocity, are presented in Fig. 11.

The FEM simulations show that when the impact velocity is equal to $v_1 = 20$ m/s then the normal stress σ_z on the impact zone reaches the quite high values up to $\sigma_z = 400$–500 MPa. This could be treated as a danger state of stress.

However, as an appropriate basic for strength assessment of impact zone, the equivalent von Mises stress is recommended. Taking this into consideration, the FEM simulations are carried out to determine also the contour line of the von Mises stress occurred in the critical zones of impact systems. Some exemplary results of these simulations for the impact zone are presented in Figs. 12 and 13.

Then, the realized values of Factor of Safety $X = \sigma_Y/\sigma_{red(max)}$ for each of these critical zones were predicted. It was assumed that the tool is made of steel with the yield strength of $\sigma_Y = 790$ MPa (this meets the strength properties of the AISI 4340 alloy steel). The relevant results are presented in Fig. 14.

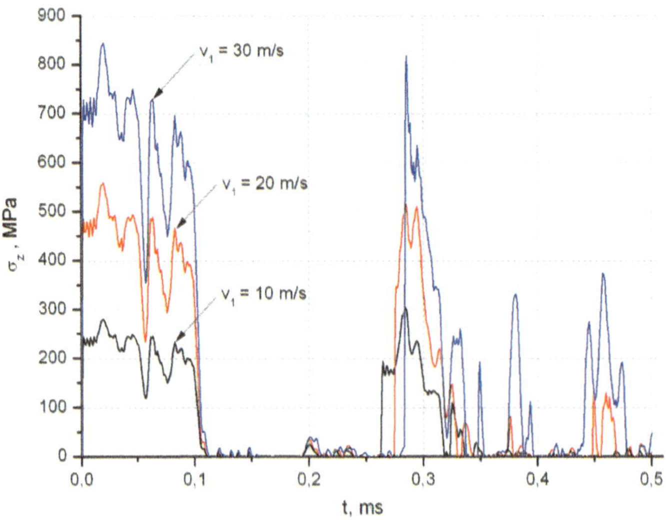

Fig. 11 Normal stress curves σ_z at the center of the contact surface "beater-tool" (v_1—impact velocity of beater)

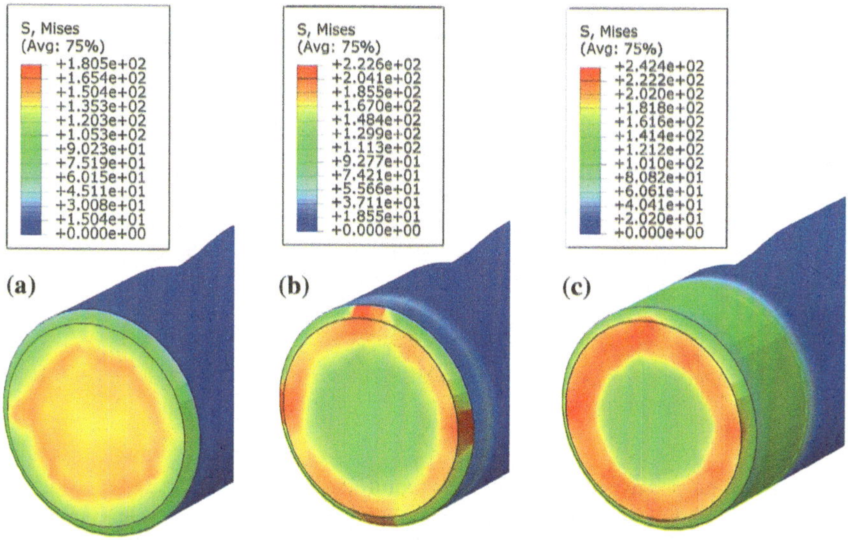

Fig. 12 Contour lines of the equivalent von Mises stress σ_{red} in the impact zone under the velocity $v_1 = 10$ m/s: **a** at the time $t = 2$ ms, **b** at the time $t = 4$ ms, **c** at the time $t = 10$ ms [20]

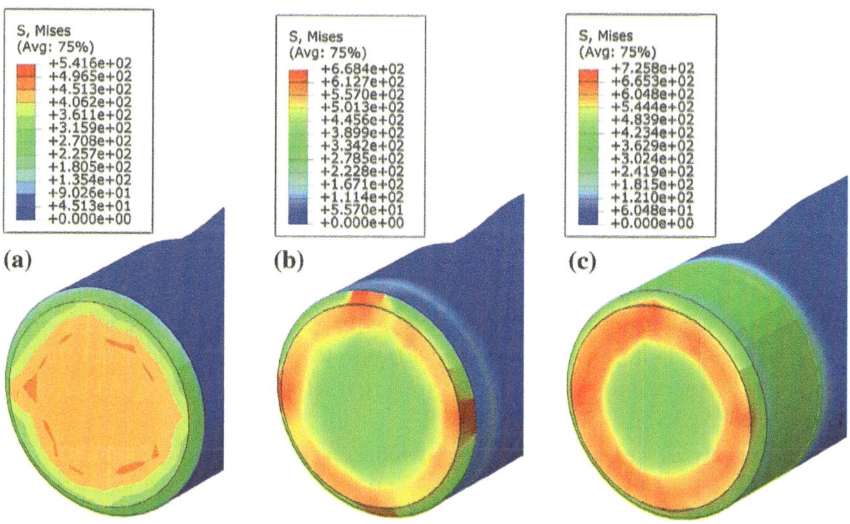

Fig. 13 Contour lines of the equivalent von Mises stress σ_{red} in the impact zone under the velocity $v_1 = 15$ m/s: **a** at the time $t = 2$ ms, **b** at the time $t = 4$ ms, **c** at the time $t = 10$ ms [20]

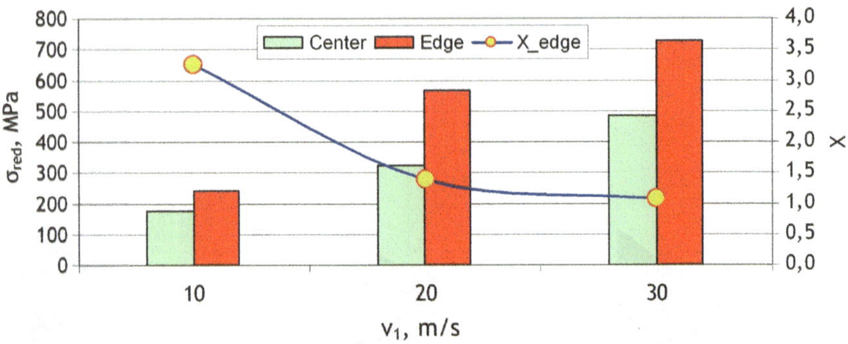

Fig. 14 Maximum values of the equivalent von Mises stress σ_{red} in the impact zone and the factor of safety X for the edge of contact surface

The simulations have shown that if the required safety factor $X \geq 1.5$ is assumed, then the permissible impact velocity v_1 should be not greater than $v_1 = 20$ m/s.

However, it should be noticed that in the real conditions the strength of tool material as well as the stresses in the critical zones of impact system are random variables. That is why the analysis of impact excitation systems of hydraulic hammers should be treated in a probabilistic approach. This can be carried out when the density functions of statistical distributions of strength and stress are known.

This probabilistic approach to strength analysis of impact systems was presented in the works [19, 20].

8 Summary

Wide range of applications and a large variety of solutions of hydraulic hammers require rational decision-making which hammer is the most appropriate in the operating conditions? In this context, the need arises to formulate indicators and characteristics of hammers and to establish objective criteria for the evaluation and selection of these devices.

In this chapter, only some important problems related to hydraulic hammers were treated and especially development trends, constructional solutions and research problems were discussed.

The main characteristics of basic operational structures and structural structures of hydraulic hammers are defined. The barriers to the further development of conventional hammers have been also formulated.

As examples, an analytical study of the hydraulic excitation system and a numerical study of the impact excitation system were presented. On this basis, the maximum permissible impact velocity in conventional hydraulic hammers was determined.

The presented material is a proverbial "tip of the iceberg". Due to the limited scope of this chapter, the problems of the analysis of residual processes, in particular vibration and noise, were omitted.

However, the residual processes (vibration and noise) are also important because they can be used for:

- the assessment of occupational hazards of operators as well as the environmental hazards,
- the purpose of diagnosing the technical condition of hydraulic hammers.

More significant problems related to hydraulic hammers were presented in details in other works of the author. This applies in particular to many years of research in laboratory and industrial conditions. These studies were conducted as a part of cooperation with centers from the copper and hard coal industries.

References

1. Hlushkova, D.B., Ryzhkov, YuV, Kostina, L.L., Demchenko, S.V.: Increase of wear resistance of the critical parts of hydraulic hammer by means of ion-plasma treatment. Voprosy Atomnoj Nauki i Tekhniki **1–113**, 208–211 (2018)
2. Hawryluk, M. Mrzygłód, B.A.: Durability analysis of forging tools for different operating conditions with application of a decision support system based on artificial neural networks (ANN). Eksploatacja i Niezawodnosc—Maint. Reliab. **19**(3), 338–348 (2017)
3. Sokolski, M.: Über die Struktur der Stosserzwingung in Arbeitssystemen von Hydraulikhämmern, [w:] „Off-road machine and vehicles in theory and practice". In: Dudziński, P. (ed) Proceedings of the 1st International conference, 23–24 Sept 1996, pp. 383–389. ISTVS-East Eur., Office Tech. Univ., Wroclaw (1996)
4. Sokolski, M.: Młoty hydrauliczne - tendencje rozwojowe w minionym ćwierćwieczu (Hydraulic hammers—development trends in the past quarter – century). Górnictwo Odkrywkowe. R. 50, nr 4/5, 123–128 (2009) (in Polish)
5. Sokolski, M.: Postawy syntezy charakterystyk młotów hydraulicznych (Fundamentals of Synthesis of Characteristics of Hydraulic Hammers). Publishing House Oficyna Wydawnicza Politechniki Wroclawskiej, Wrocław (2013). (in Polish)
6. Sokolski, M.: Neue Entwicklungsrichtung bei Hydraulikhämmern für Erdbaumaschinen, [w:] „Mechanisierung im Erdbau". XIII. Internationale Konferenz, TU Dresden, 6–8 Sept 1988. Dresden 1988, 1–4
7. Bauer, E.E.: Contech Mining Application Manual. Consolidated Technologies. Corp., Mining Division, Denver (sine anno)
8. Grantmyre, I.: Hawkes I.: High-energy impact rockbreaking. Can. Min. Metall. Bull. (1975)
9. Etherington, M., Deering, D.: The fluid tappet—a new concept in high energy impact breaking. In: Earthmoving Industry Conference, Central Illinois Section, Peoria, IL, 10–12 April 1978
10. Zhu, M., Zhao, S., Dong, P.: Modelling and simulation of a direct drive reversing valve for the hydraulic die forging hammer. In: 2016 IEEE International Conference on Aircraft Utility Systems (AUS), Beijing, pp. 206–211 (2016)
11. Redelin, R.A., Kravchenko, V.A., Kamanin, Y.N., Panichkin, A.V., Bozhanov, A.A.: Study of effect of in-line hydropneumatic accumulators on output characteristics of hydraulic hammer. In: IOP Conference Series: Earth and Environmental Science, 10/2017, Vol. 87, Issue 2, pp. 22016, ISSN 1755–1307
12. Krauze, K.: Experimental determination of impact energy of hydraulic hammers. Arch. Mech. Eng. **XLVII**(1), 21–32 (2000)

13. Szykowny, K.: Określenie energii udaru młotów hydraulicznych na podstawie pomiarów wybranych parametrów pracy (Determination of impact energy of hydraulic hammers on the basis of measurements of operation parameters), (PhD Thesis, supervisor: prof. K. Krauze), AGH Kraków – Lubin (2006) (in Polish)
14. Boczarow, J.A., Prokofiew, W.N.: Napędy hydrauliczne pras i młotów (Hydraulic drive systems of Presses and Hammers). Publishing House WNT, Warszawa (1971) (in Polish)
15. Mezentsev, I.V.: Influence of design factors on the efficiency of hydraulic hammers. J. Min. Sci. **39**(4), 400–404 (2003)
16. Park, J.-W., Kim, H.-E.: Development of the test system for measuring the impact energy of a hydraulic breaker. In: Proceedings of the 6th JFPS International Symposium on Fluid Power, pp. 75–79, TSUKUBA 7–10 Nov 2005
17. Giuffrida, A., Laforgia, D.: Modelling and simulation of a hydraulic breaker. Int. J. Fluid Power **6**(2), 47–56 (2005)
18. Sokolski, M.: A computer-aided method for diagnostic testing of single-acting hydraulic hammers. Model. Simul. Control **B 36**(2), 55–63 (1991)
19. Sokolski, M., Sokolski, P.: Assessment of the probability of failure-free operation of the working system of a small-dimension hydraulic hammer—a case study. In: Podofillini, L., et al. (eds.) Safety and Reliability of Complex Engineered Systems, pp. 4213–4217. Taylor & Francis Group, London (2015)
20. Sokolski, M., Sokolski, P.M.: Strength estimation of the impact zone—a critical area of the tools of the hydraulic hammers. Arch. Civil Mech. Eng. **16**(4), 767–776 (2016)

Development and Parameter Justification of Vibroscreen Feed Elements

Mikhail Doudkin, Alina Kim, Marek Młyńczak, Gennady Kustarev and Vadim Kim

Abstract The article presents research results of the screening process of bulk materials with a new vibroscreen design with additional feed elements, moreover a three-dimensional solid-state computational model, as well as the results of stress-strain state analysis of feed elements rods of vibroscreen. Changes of the traditional screen structure are shown; the use of them increases the screening process efficiency and passage intensity of lower grade of bulk material to the sieve. It is established that the most simple, accessible, effective and efficient way to increase the effectiveness of the screening is the use of additional feed elements. According to the developed mathematical model of cell screening process, numerical studies to identify the effect of process parameters on the screening kinetics carried out. It is proved that the use of additional feed elements has a significant effect on the screening process kinetics and the state evolution of fine particle concentration in the bulk layer. It increases the screening effectiveness by improving the sieve filling and reduces the screening time. It is proved the performance, possibility to use, benefits, efficiency, and prospects for further research of the new vibroscreen scheme with feed elements. Description and results of comparative experimental studies of vibroscreens with different variants of additional bulk material excitation are presented. An algorithm for solving the problem numerically using the finite element method is proposed. The obtained results were used at the designing stage of the platform with feed elements for industrial vibroscreen and subsequently confirmed in work in a real experiment. The stress–strain state of feed elements rods were analyzed for various bulk materials, conditionally designated A and B, sorted by vibrating screen, where feed elements were mounted. These materials, in screening process with varying strength, acted on

M. Doudkin · A. Kim · V. Kim
Faculty of Mechanical Engineering, East Kazakhstan State Technical University, A. Protozanova Str. 69, 070004 Ust-Kamenogorsk, Kazakhstan

M. Młyńczak (✉)
Faculty of Mechanical Engineering, Wroclaw University of Science and Technology, ul. Wybrzeze Wyspianskiego 27, 50-370 Wroclaw, Poland
e-mail: marek.mlynczak@pwr.edu.pl

G. Kustarev
Faculty of Road and Technological Machinery, Moscow Automobile and Road Construction State Technical University (MADI), Leningradskiy Prospekt 64, 125319 Moscow, Russia

M. Sokolski (ed.), *Mining Machines and Earth-Moving Equipment*,
https://doi.org/10.1007/978-3-030-25478-0_10

feed elements rods, the parameters of which did not change for the flow of various bulk materials. Rods perceived this load pressing in different ways, which was shown by the finite element analysis.

Keywords Bulk material · Screening process · Vibroscreen · Feed elements

Conventional Designations

n The material layer under the sieve
Δx The thickness of the elementary layer
h The height of material (m)
p_u The probability of particles transition up to
p_d The probability of particles transition down to
p_s The probability of particles stay at the cell
i Cell number
k Number of transition
Δt Time of transition from one state (cell) only to another, but not is possible further (s)
t_s Time for passing of all layer of material to the sieve (s)
t_k Current moments of time, i.e. the actual finding of a particle in space (s)
P Transition probabilities matrix
v_i Convective component
V_i Dimensional segregation speed (m/s)
D_i Dimensional dispersion coefficient (macrodiffusion factor) (m^2/s)
d_i Symmetrical purely random component with zero expectation
v_f The probability of particles penetration through the sieve
C_i The probabilities of the fact, that at any given time the particle will be in i-cell
E_i^k State of the cells chain

1 Introduction

Screening is a process of separating the bulk particulate material to size classes by means of screening it through one or more series or parallel screening surfaces with orifices [1, 2].

As a result of the screening process, the material remaining on the sieve is produced-oversize fraction or minus mesh. The material supplied to the screen is called initial, and screening products are called classes. The class that is used in the production as a readily marketable product is called grade [3–5]. Separation processes are important in industry. The understanding of the fundamentals of granular separation on sieve surfaces is incomplete [6].

Feed elements are an additional construction that significantly intensifies the screening process and at the same time the weakest part of the vibroscreen. Based on the developed mathematical screening model using a main platform of feed elements, numerical studies have been carried out to identify the influence of the process parameters on the screening kinetics and the state of various passage particles in the screened layer. The numerical studies have shown that the introduction of additional feed elements into the screened material flow leads to an increase in the passing intensity of the bulk material lower class to the sieve, regardless of the size fractions of different bulk materials.

Figure 1 schematically shows a vibrator with additional feed elements (FE) 7 mounted on the frame 6.

During the screening process, the sorted material flow 2 moves along the screen 1 and comes into contact with the feed elements (FE) 7 located along and across the screen. Since the feed elements themselves are installed with a certain step, the material not only rests against them but also flows around the sides, exerting pressure on them with a certain force for each material.

The collision of bulk material particles with FE leads to the creation of additional chaotic movements of particles in the upper layers of bulk material, either coinciding or different from the forced oscillations direction from the vibration exciter 5 (Fig. 1). While the difference in particles velocities of the particulate material relative to each

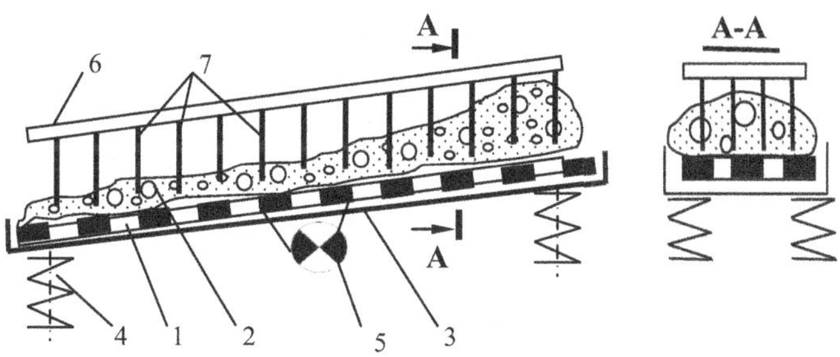

Fig. 1 Scheme of vibroscreen with additional feed elements: 1—screen, 2—bulk material, 3—box, 4—elastic supports, 5—vibration engine, 6—frame, 7—feed elements

other, which activates the passing particles process of the bottom fraction to the screen surface through the entire layer of bulk material through forced mixing.

2 Concept of Innovative Vibroscreen

The separation process of particles is called sieving and technological machines equipped with sieve baskets—screens. In case of screening of large particle size (with low specific surface) the layer of material at the screen can be represented by one or more particles of average size. All the particles are located in the immediate vicinity of the sieve, and the efficiency of passage of particles depends on the ratio of their size and the sieve orifices size [7–14].

In case of smaller particles separation (with high specific surface) the distribution of the material in a thin layer along the screening surface is unacceptable due to high required screen surface. The material is supplied to the screen by a relatively thick layer. In this case, passage particle requires some time to reach the screening surface. This time can be a determining factor in the kinetics formation of passage particles extracting, i.e. determines the screening productivity. In addition, for the particles movement to the screening surface, it is necessary to ensure their mobility in the layer, that is, to bring the material to a pseudo-fluidized state, for which vibration is used for the material from the screening surface. Thus, the physical content of this process is random migration of particles in vibro-fluidized material layer with the possibility of exit of passage particles to the screening surface. The full exit of all passage particles to the screening surface determines the kinetics of screening [15–20]. In other words, screening process is divided into two phases: passing of material particles of the lower class to the sieve through all layer, and passing of material particles through the sieve. Increase of dynamic impact on bulk material considerably improves the first phase of screening process. Consequently, productivity of screening process rises. The proposed technical solution is directed to the acceleration of the first phase of screening process.

As a result of the studies, new vibroscreen was developed and manufactured. The vibroscreen containing a box with a sieve mounted on elastic supports, vibration generator and feed elements (FEs) installed over the screen surface, made in the form of bars fixed to the frame; feed elements may be movable and provided with a drive, as well as fixed and detachable (Fig. 2). This solution is patented [21].

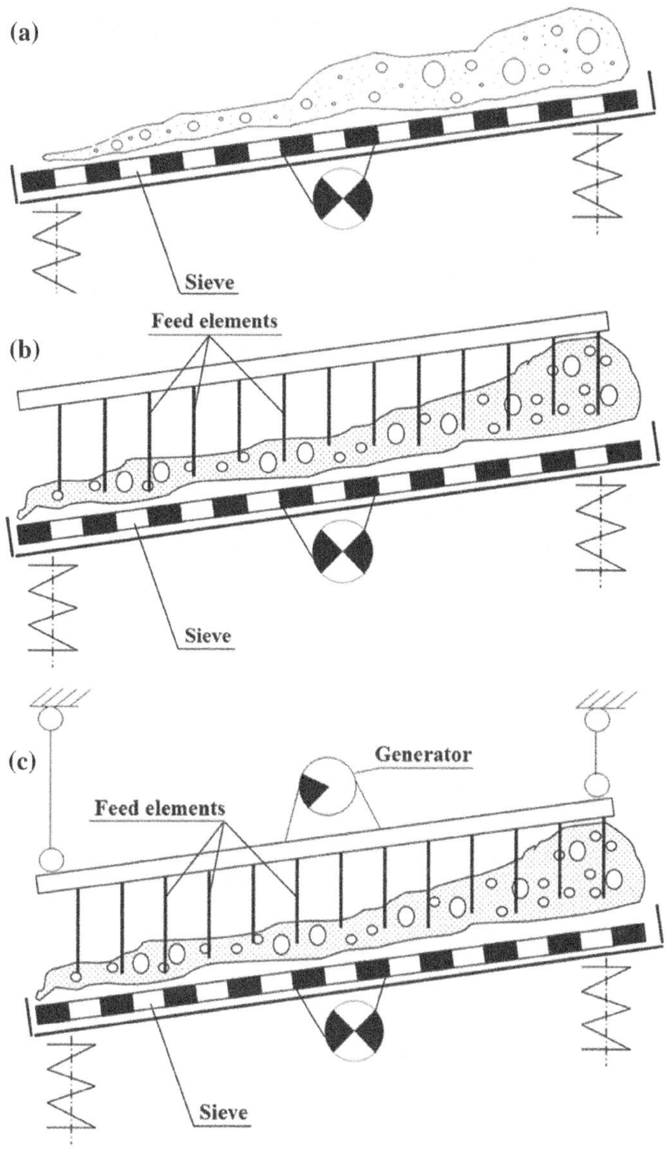

Fig. 2 Schematic view of vibroscreens: **a** Vibroscreen without with feed elements (FEs), **b** vibroscreen with static feed elements (sFEs), **c** vibroscreen with dynamic feed elements (dFEs)

3 The Mathematical Model of the Separation Process

In case of absence of feed elements, i.e. in a normal vibroscreen, in a course of movement, particles of the lower class make migration through a material flow to the screening surface. At the same time for detailed reviewing of behaviour of lower class particles, all volume of material is divided into elementary layers. Thickness of one elementary layer can have different values. We suggest considering layer thickness equal to the maximum size of a passage particle. For simplification, all particles are presented in the form of spheres of a certain diameter. In a normal vibroscreen on the speed of particles passing through all volume of material to the sieve, amplitude, and frequency of screening surface influence generally. In a normal vibroscreen, we practically cannot accelerate the first phase of screening. Our decision is intended for an intensification of the first phase of screening.

Let us consider the physical picture occurring on a sieve. When transporting material, particles of the lower and top classes move to an exit (in the direction of vibration). Besides, particles make the movements in a material layer. At the same time, there are processes of segregation and diffusion. The behaviour of separate particle of material is described in many sources [2, 15] and represents chaotic movement in all possible directions. The randomness arises from impacts of particles among themselves (Fig. 3).

With the using of feed elements, particles of material begin to collide, except collision with each other; they also collide with feed elements (Fig. 4). This could theoretically increase their dynamic activity, particles will quickly move to the screening surface, i.e. the first phase of screening is accelerated.

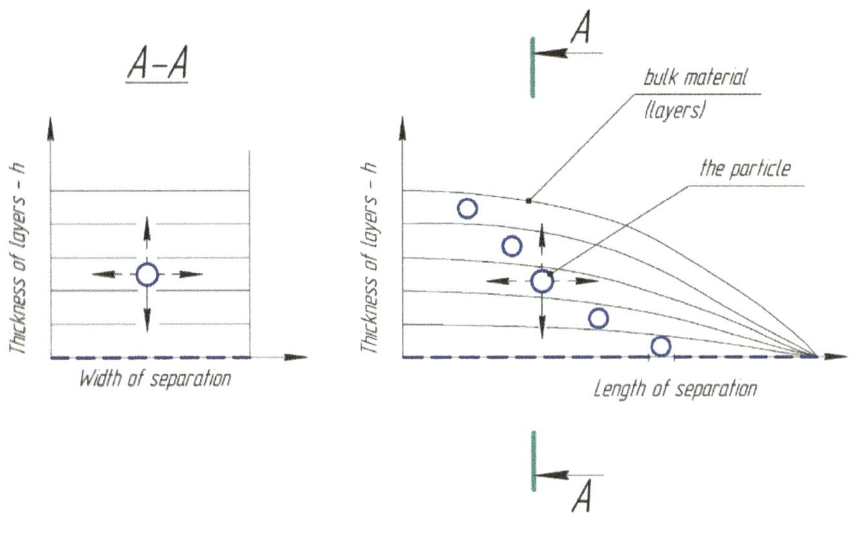

Fig. 3 Separation process on the vibroscreen

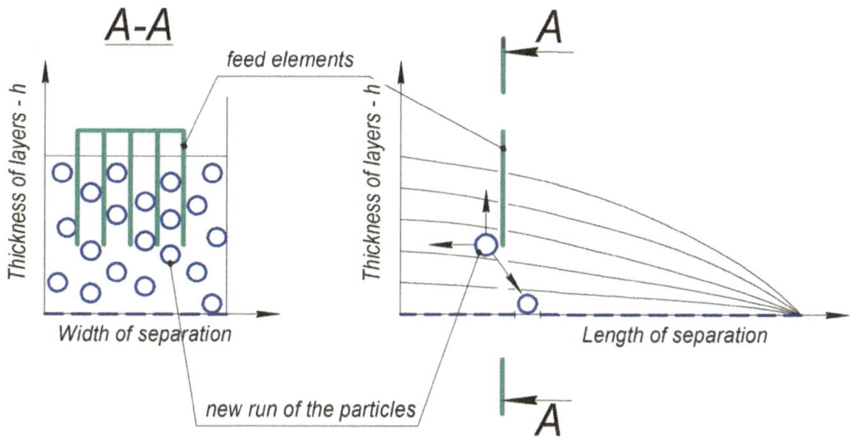

Fig. 4 Separation process with static (sFEs) and dynamic feed elements (dFEs)

The whole layer of material is divided into $n-1$ elementary layers of final size, where n—the material layer which is under the sieve. Inside of each layer, the distribution of all parameters of the material over its volume is considered uniform. The thickness of the elementary layer Δx must be greater than the size of the largest particle, but significantly less than the total layer thickness. The material layers are called cells of our probabilistic model, thus we obtain the well-known cell model representation of the screening process [15]. The total number of cells chains determines the total space of particle states. Figure 5a shows the probability of particles

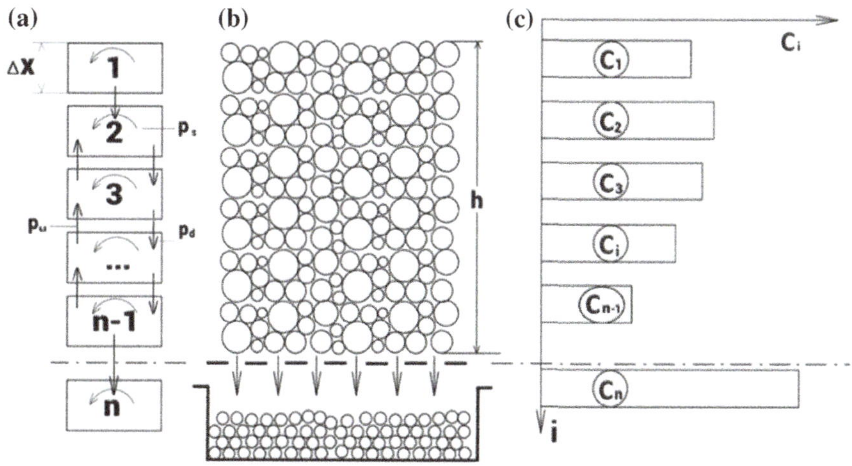

Fig. 5 Cell model of the process and the allocation scheme in the transition probabilities of the symmetrical component (**a**), the estimated process scheme (**b**), an example of the probability of distribution of the states (**c**)

transition up to p_u, down to pd and stay at the cell p_s.

In Fig. 5b actual state of material by h height on the sieve is shown. In Fig. 5c, distribution of probabilities of particle finding of the lower class in a concrete elementary layer (cell) is considered. Let us consider this scheme in more detail.

The probabilities of C_i, state that at any given time the particle will be in i-cell are different (Fig. 5c). The complete set of them forms a state vector of particle of the material.

$$C = \begin{bmatrix} C_1 \\ C_2 \\ \cdots \\ C_n \end{bmatrix}, \tag{1}$$

sum of all elements of which is equal to one.

Let us consider the process through consecutive small periods Δt—time of transition from one state (cell) only to another, but not is possible further. Time for passing of all layer of material h height to the sieve will be equal to $t_s = \Delta t \bullet (k - 2)$ where integer $k = 1, 2, \ldots$ (transition number of one cell to next) becomes an integer analogue of time t_s. Then current moments of time, i.e. the actual finding of a particle in space, will be calculated as $t_k = (k - 1) \bullet \Delta t$ [16].

Most of the time t_s obviously consumed by the first phase of screening, i.e. passing the material to the screening surface. By reducing this time, we can accelerate the In this case we can improve the efficiency of the screening process. In our work, effectiveness of screening process is meant overall efficiency of the screening process and not clogging the sieve (classical efficiency). Thus, the acceleration of the first screening phase occurs by accelerating the time passage of material layer to the sieve ts, which is associated with the state of cells chain E_i^k.

The state of the cells chain can be described by the following matrix recurrence equation:

$$\boldsymbol{E}_i^k = \boldsymbol{P} \cdot \boldsymbol{C}^k \tag{2}$$

where: $\boldsymbol{P}$ is the transition probabilities matrix, which with current restrictions for Δt will have the following view:

$$\boldsymbol{P} = \begin{vmatrix} p_{s1} & p_{u2} & 0 & 0 & 0 & 0 \\ p_{d1} & p_{s2} & p_{u3} & 0 & 0 & 0 \\ 0 & p_{d2} & p_{u4} & p_{u4} & 0 & 0 \\ 0 & 0 & p_{d5} & p_{s4} & p_{u5} & 0 \\ 0 & 0 & 0 & p_{d4} & p_{s5} & 0 \\ 0 & 0 & 0 & 0 & p_{d5} & 1 \end{vmatrix} \tag{3}$$

This matrix in the column corresponding to the cell number i ($i = 6$), contains the probabilities of transition during Δt up to p_{ui}, down to p_{di} and stay at p_{si}. The sum of all the probabilities in each column is equal to one.

In general, the probability to go up and down is not equal. For example, in case of tendency of the fine particles for segregation downwards, the probability will be $p_{ui} < p_{di}$. Symmetric purely random component with zero expectation d_i and convective component v_i is related to the parameters of the dispersion equation by the relations:

$$v_i = \frac{V_i \Delta t}{\Delta x}, \tag{4}$$

$$d_i = \frac{D_i \Delta t}{\Delta x^2}, \tag{5}$$

where:

V_i dimensional segregation speed,
D_i dimensional dispersion coefficient (macrodiffusion factor).

At the screening on the usual screen (without feed elements), there is segregation and diffusion, but it can be assumed that the using of feed elements, these parameters need to be changed. Moreover, the diffusion coefficient is changed in a positive way, i.e. it accelerates the process of passing the particles to the sieve. It is obvious that these parameters can vary depending on the type of feed elements, their installation, number, etc. Taking into account the aforesaid, the transition probability matrix (3) is the mathematical representation of the layer of material on the sieve.

A special place is given to the cell corresponding to the material layer which is under the sieve. Passage particles trapped in it remain there forever, so $p_{sn} = 1$. The probability of an exit to that a cell from the bottom cell, belonging to the layer of material, is different from other downwards transition probabilities:

$$v_f = p_{dn} - 1 \tag{6}$$

This probability is largely dependent on the ratio of passage particle sizes and sieve orifice. The matrix (3) recorded using new data has the following form:

$$\boldsymbol{P} = \begin{vmatrix} p_{s1} & p_{u2} & 0 & 0 & 0 & 0 \\ p_{d1} & p_{s2} & p_{u3} & 0 & 0 & 0 \\ 0 & p_{d2} & p_{u4} & p_{u4} & 0 & 0 \\ 0 & 0 & p_{d5} & p_{s4} & p_{u5} & 0 \\ 0 & 0 & 0 & p_{d4} & p_{s5} & 0 \\ 0 & 0 & 0 & 0 & p_{d5} & 1 \end{vmatrix} \tag{7}$$

where v and d values are taken the same for all layers of cells and the elements of main diagonal are calculated as the difference between one and the sum of all other probabilities in the column.

If the initial probabilities distribution is known, identical to the initial distribution of the relative concentration of passage particles of the considered material fraction, the Eq. (2) in case of a known matrix (7) completely describes the kinetics of the process. Considered this model we can come to positive or negative conclusion about the efficiency of use of feed elements, i.e. about their influence on performance and efficiency of screening process in general.

Based on this model (2) we have carried out computer calculations which have shown dependence of probability of time acceleration of particles passing to the sieve. All calculations have been carried out for experimental physical model of a new vibroscreen (see Sect. 4). In Fig. 8 three cases are presented: without feed elements (FEs) (Fig. 6a), with static feed elements (sFEs) (Fig. 6b), and dynamic feed elements (dFEs) (Fig. 6c).

In Fig. 6a is shown that in normal screen (without feed elements) particles of material of the lower class reach the screening surface in 30 s. When using feed elements, time of particles passing of the lower class through a layer of material accelerates for 6.66% and constitutes 28 s (Fig. 6b). When using dynamic feed elements, time of passing of material to the sieve constitutes 26 s, i.e. the first phase of screening accelerates for 13.33% (Fig. 6c). Thus, it is possible to assume reasonably that with use of feed elements, screening efficiency increases for 6.66–13.33%.

The productivity of screening before FE can be described by the formula [1]:

$$Q = \mu_{\mathrm{BE}} \cdot (S - S_{\mathrm{BE}}) \cdot \sqrt{\frac{2}{\gamma - \Delta\gamma}} \left(\frac{m(t) \cdot \left(0.02 \cdot A \cdot \vartheta^2\right)}{S} - \frac{k_{\mathrm{TP}} \cdot \gamma(t) \cdot c \cdot \left(2.22 \cdot A \cdot \vartheta^2\right)}{2} \right) \quad (8)$$

It can be seen from the formula (8) that the increase in the productivity of the screening process is possible due to decrease in the specific material mass on the screen or by more intensive mixing of the bulk material on the screen, which leads to slowing down of the material flow, as well as its redistribution and longer residence of the material in the active sieving zone. As a result, the specific weight of the finished product increases, the total amount of material decreases, while the productivity Q increases; because smaller amount of material is distributed over the same area of material flow and experiences less resistance.

Since active screening occurs mainly up to FE, after them the vibrating screen can be considered as a vibration transporter, since the screening process after FE itself is very insignificant, so that the productivity of the process after FE is directly and primarily related to the speed of material transportation along the screen $Q = v(t)$, and velocity of the material passing through the screen is the main researched parameter after the FE [1].

$$\frac{v_2^2}{v^2} = \frac{2m \cdot A_2 \cdot \vartheta^2 \cdot \phi_2^2}{S_2 \cdot (\gamma - \Delta\gamma)} \cdot \frac{S \cdot \gamma(t)}{2m \cdot A \cdot \vartheta^2 \cdot \phi^2} \Rightarrow \frac{v_2^2}{v^2} = \frac{A_2 \cdot \phi_2^2 \cdot \gamma \cdot S}{A \cdot \phi^2 \cdot (\gamma - \Delta\gamma) \cdot S_2} \quad (9)$$

According to the formula (9), it can be concluded that the rate of bulk material after FE should increase for the following reasons:

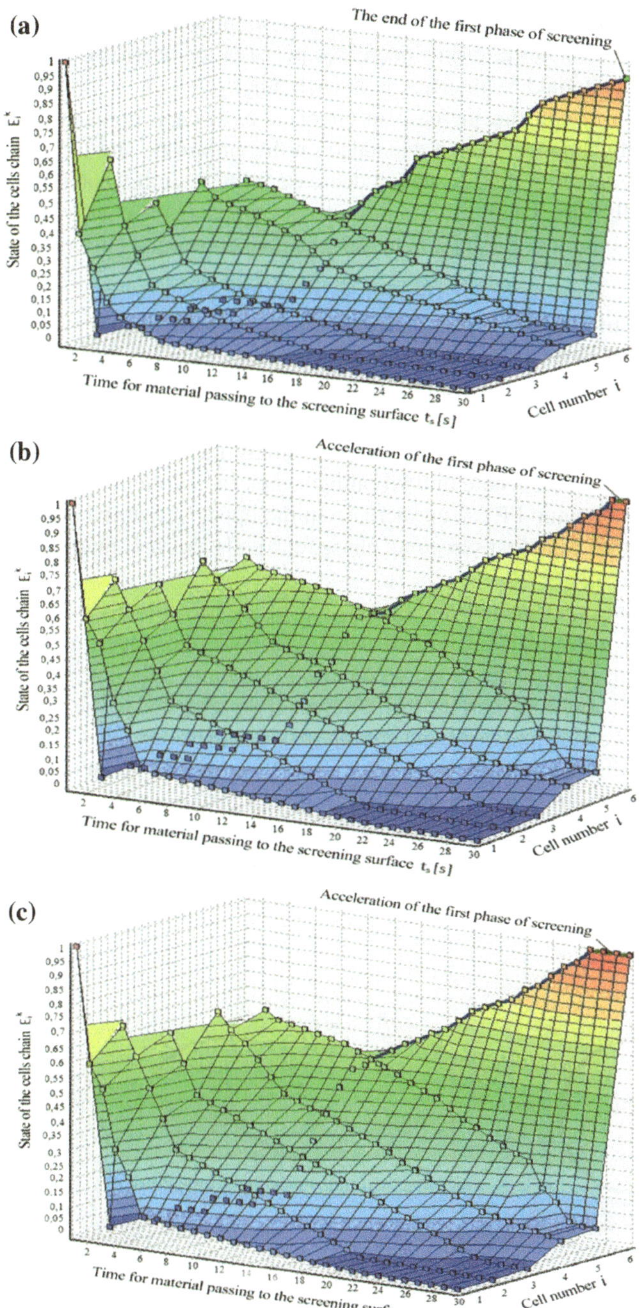

Fig. 6 State of the cells chain of lower class of the vibroscreen: **a** without with feed elements (FEs), **b** with static feed elements (sFEs), **c** with dynamic feed elements (dFEs)

- due to the decrease in the material resistance to movement, because, after the FE, the number of material pieces decreases per area, respectively, the pieces of the remaining material less interfere with each other;
- in connection with the increase in energy, i.e. at each point, a certain energy was initially supplied, which depended on the amplitude and frequency of sieve vibrations, and after the FE, with a decrease in the specific material mass, which was partially sorted and fell into the subsystem space, the energy that was set for the entire volume of the bulk material was now consumed only on his remaining part.

After substituting the experimental data into the formula (9), mathematical calculations confirmed the tendency of first screening phase acceleration, i.e. the time of material passage to the screen is accelerated by the use of additional FEs at 6.66% with the use of static FE and by 13.33%, with the use of dynamic FE, the overall efficiency of the screening process is accordingly increased, which is graphically depicted in Fig. 7, each state of a vibrating screen (a—without a FE; b—with static RE; c—with dynamic FE) as an increase in the upper horizontal coordinate shows the acceleration of the first screening phase for 6 cells, conventionally accepted lying on the screen.

To carry out experimental studies and analyze the behaviour of particles on a natural flat screen, a physical model of vibrating screen was developed and manufactured, which contains a sieve box mounted on elastic supports, vibrator and installed feed elements (FE) above the surface of the screen, rods attached to the frame, while the feed elements can be movable and equipped with a drive, and also fixed removable.

These requirements, even before the production of feed elements complex, lead to the need for setting new tasks and developing new modelling techniques at the stage of feed elements designing for vibration screens that most fully and adequately take into account the geometry, internal composition and mechanical features of the elements being developed for operation under real operating conditions.

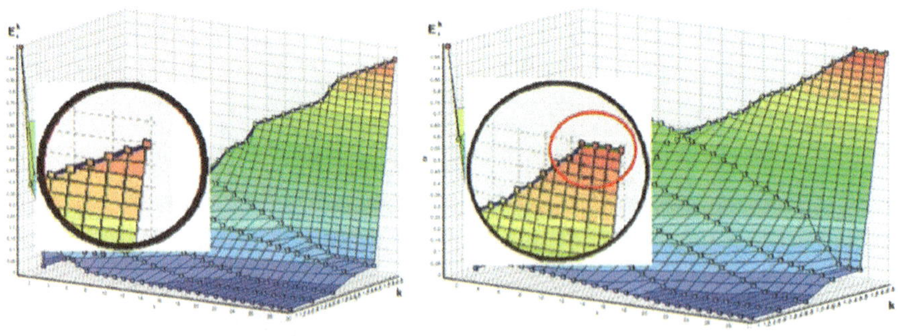

Fig. 7 Graphic image of the acceleration of the first screening phase of the conditional 6th cell: **a** conventional vibroscreen (without FEs); **b** vibroscreen with static FEs; **c** vibroscreen with dynamic FEs, according to the developed mathematical model

From these conditions, higher requirements to the rigidity and strength of the feed elements arise, while simultaneously striving to reduce their mass, optimality and versatility of the design, while maintaining the reliability of the work and the FE, and the vibroscreen itself.

One of the main methods that reduce the structure development time, which increasing requirements for rigidity and strength characteristics, is the finite element method (FEM), which makes it possible to foresee the behaviour and reliability of the experimental design, even before industrial manufacture. This makes it possible to shorten the cycles of experimental studies, modification of the design, processing of design and technological documentation, significant reduction in the cost of production preparation.

Therefore, the analysis of design decisions in the early design stages with the help of simulation systems, including the finite element analysis, which allows modelling the future design and the processes of its testing for various effects at the design stage, acquires special significance.

In this article, the finite element method (FEM) is used to solve the problems of determining the stress–strain state of feed elements rods fixed on the platform, for which we will perform the necessary preliminary graphic constructions, calculations and selection of initial data.

A solid 3D model of a vibrating screen with a separate bearing surface of feed elements rods (Fig. 8) and separate calculation models (Figs. 8, 9, 10, 11 and 12) of feed elements consisting of metal rods mounted on a bearing platform from metal corners were developed and implemented in a system of parametric solid modelling KOMPAS-3DV17.1. The calculations were made in the APM FEM system, which is an integrated tool in the KOMPAS-3DV17.1 for the preparation and subsequent finite element analysis of a three-dimensional solid model (Certification passport No. 330 dated April 18, 2013, issued by the Federal Service for Environmental, Technological and Nuclear Supervision (Rostekhnadzor), FBU "STC NRS".) The calculating core

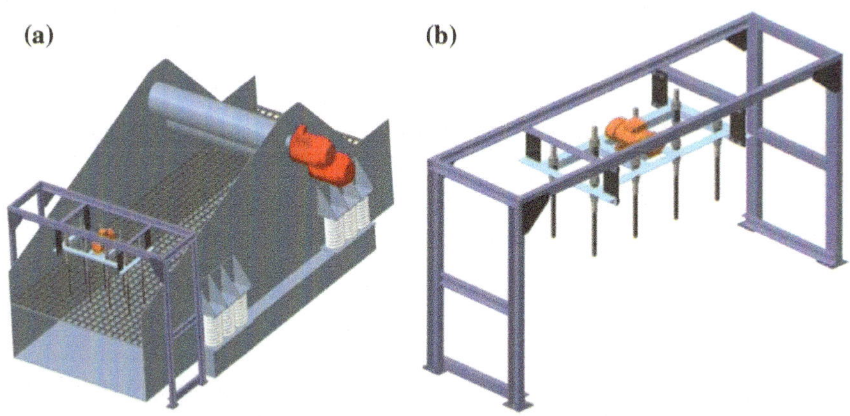

Fig. 8 Vibroscreen with feed elements: **a** general view of vibrating screen with FE; **b** feed elements

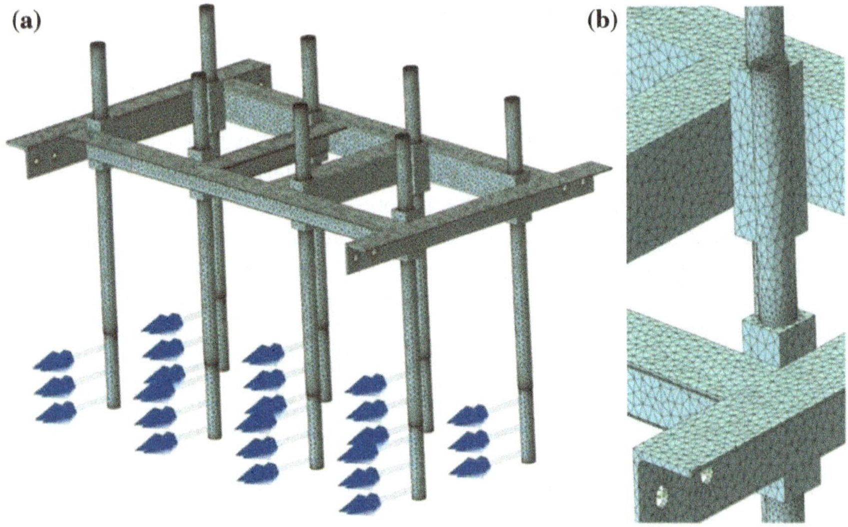

Fig. 9 Finite element grid of the FE construction (**a**) and its fragment (enlarged) (**b**), consisting of tetrahedron grid of finite elements

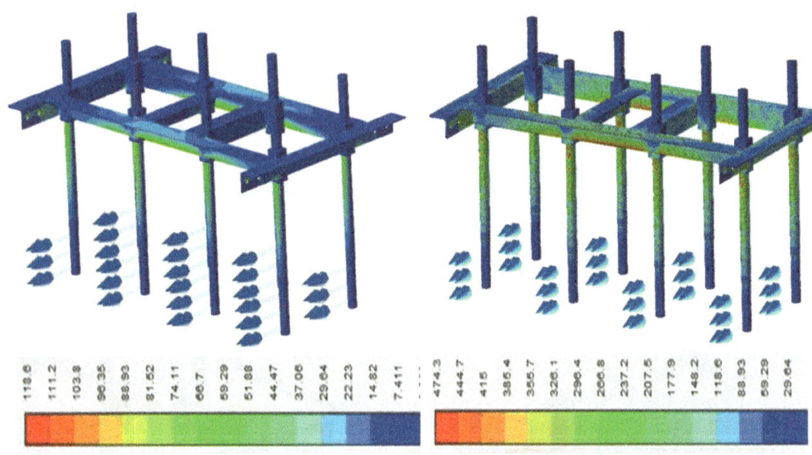

Fig. 10 Equivalent Mises stress of feed elements platform for various bulk materials

of the APM FEM system for KOMPAS-3D is the “Finite element program system APM Structure 3D”.

In the considered structure of the vibroscreen, the investigated element will be a system of feed elements consisting of a bearing surface and eight vertical rods arranged 4 in two rows and at a certain distance from each other (Fig. 8).

The setting of the feed elements on the vibrating screen is shown in Fig. 3. The calculated finite element grid of the platform with feed elements is shown in Fig. 8.

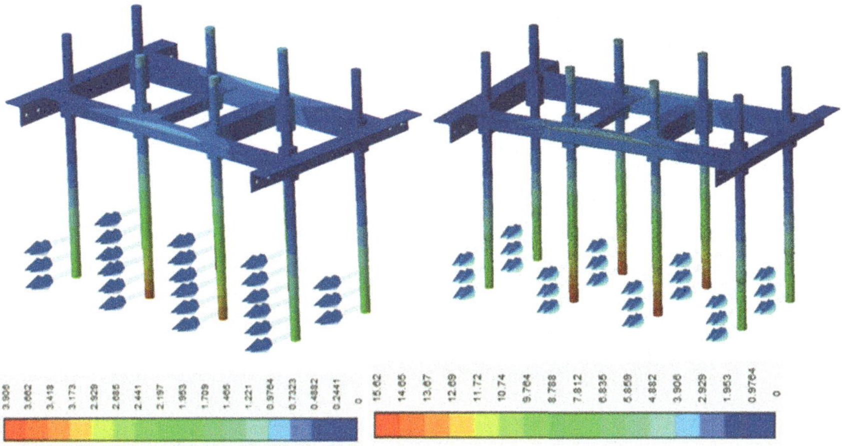

Fig. 11 Distribution of the total linear displacement on the platform

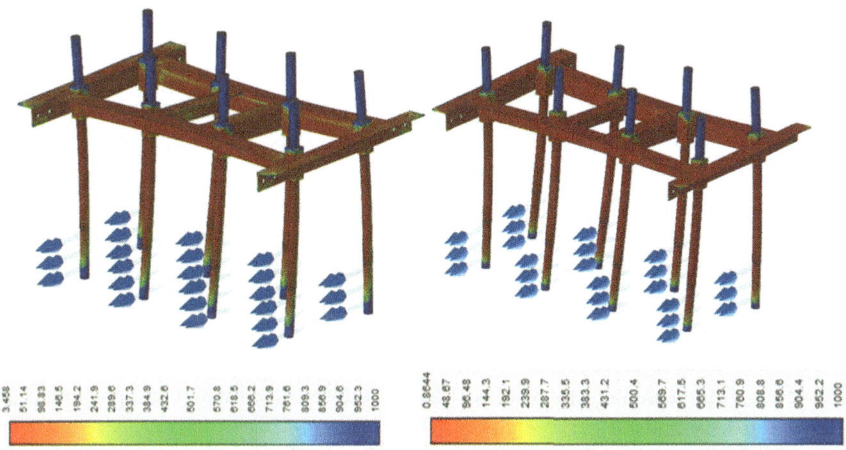

Fig. 12 Distribution of the safety factor for strength on the platform

Between feed elements by conditional groups, the sorted material with a shadow trace from the motion is shown schematically.

Conditionally it is considered that the bulk material flow during the passage of the feed elements rods is the same in height both in the middle of the screen and at edges, which is shown in Figs. 9, 10, 11 and 12.

The structure of the feed elements can be conditionally divided into two groups of homogeneous elements—a metal corner (Steel 20, Table 1), from which the bearing platform is made (Corner profile d63 × 63 × 5 GOST 8509-93, transverse length $L = 1066$ mm and longitudinal length $L = 340$ mm), which serves as the basis for

Table 1 Data of platform material (Steel 20, GOST 1050-2013)

S. No.	Parameter	Value
1	Yield stress (MPa)	235
2	The modulus of normal elasticity (MPa)	200,000
3	Poisson's ratio	0.3
4	Density (kg/m^3)	7.800
5	The temperature coefficient of linear expansion (1/°C)	0.000012
6	Thermal conductivity (W/m °C)	55
7	Compressive strength (MPa)	410
8	Tensile strength (MPa)	209
9	Torsional fatigue strength (MPa)	139

fastening the rods, and the rods themselves (Steel 45, GOST 1050-2013), mounted on this site.

The behaviour of feed elements is simulated in contact with two different bulk materials "A" and "B", which have significant property differences. For each distributed force acting on feed elements, the corresponding objects are selected and the following load parameters are applied: Force vector: for material "A"—$X = 400$; $Y = 0$; $Z = 0$; for material "B"—$X = 1200$; $Y = 0$; $Z = 0$. Accordingly, the magnitude of the flow force of the bulk material acting on the rods of the feed elements: for material "A"—400 N, for material "B"—1200 N.

Accordingly, information on fixations and coinciding surfaces is included in the program, and finite-element grid consisting of 10 nodal tetrahedrons is chosen. It should be noted that tetrahedrons are the only elements that can be used to thicken the adaptive grid. When the density of the grid is increased, the solution becomes more precise. So it can unerringly assume relatively slow change in the stresses in these areas. In these calculations, the adaptive grid is unchanged. For all accepted parameters, static calculations were performed, the results of which are presented in Figs. 8, 9, 10, 11 and 12.

The initial data for solving the problem are summarized in Tables 2, 3.

The equivalent Mises stress (Fig. 10), SVM (MPa), ranged from the minimum value (0) to the maximum (118,579,194) for bulk material "A" and from 0 to 474,317,501 for material "B" (Table 4).

Stress analysis shows their growth for material "B", which is very clearly seen when comparing the FE in Fig. 10, where the middle of the area working with material "B" acquired tints of stresses corresponding to high numerical values of the colour scale.

Accordingly, the total linear displacement, USUM (mm), also varied from (0) to a maximum of 3.905783 for bulk material "A" and from 0 to 15.623117 for material "B" (Fig. 11).

As a result of the numerical analysis of the researched system, a picture of the safety factor distribution by structural elements was obtained, expressed by the range

Table 2 Data of feed elements material (Steel 45, GOST 1050-2013)

S. No.	Parameter	Value
1	Yield stress (MPa)	560
2	The modulus of normal elasticity (MPa)	210,000
3	Poisson's ratio	0.3
4	Density (kg/m^3)	7.810
5	The temperature coefficient of linear expansion (1/°C)	0.000013
6	Thermal conductivity (W/m °C]	47
7	Compressive strength (MPa)	600
8	Tensile strength (MPa)	294
9	Torsional fatigue strength (MPa)	150

Table 3 Parameters and results of feed element location

S. No.	Description	Value
1	Element type	10-nodal tetrahedrons
2	Maximum length of the element side (mm)	10
3	The maximum coefficient of condensation on the surface	5
4	The coefficient of vacuum in the volume	3.5
5	Number of finite elements	114,953
6	Number of nodes	235,487

of its change. The obtained values of the safety factor ranged from 3.457605 to 1000 for bulk material "A" and from 0.8644 to 1000 for material "B" (Fig. 12). As can be seen from the results of the analysis, the loading regime of the structure for material "A" is safe, as indicated by the value of the minimum safety factor of 3.5. And for material "B", the minimum safety factor is 0.8, which is much less than 3.5. Consequently, the load from the material "A" practically does not affect the efficiency of the FE. The simulated design demonstrated a large safety margin, confirmed by the resulting safety factor.

The scale of the FE image shown in Fig. 12 does not allow us to clearly compare the colour variations of the finite element grid, which confirm changes in the calculated values of the coefficient for the loads considered from the action of various materials, but the colour gradient decoding below each of the figures shows numerically a significant difference in the indices strength.

Based on the results of modelling and calculation, there were no large deformations or stresses in the structure exceeding the yield strength of the material when the two materials "A" and "B" were screened, so there was no point in resolving the problem with the new data.

An exception can be given by a material with special properties that increases the force of the pressure on feed elements an order of magnitude because of its increased density or the adhesion coefficient. The above calculations and modelling

Table 4 Inertial characteristics of FE models

S. No.	Name	Value
1	Weight of the model (kg)	55.810975
2	Centre of the model gravity (m)	(0.000011; −0.117649; −0.000087)
3	The inertia moment of the model relative to the mass centre (kg m^2)	(8.229942; 7.831576; 3.248618)
4	The reactive moment with respect to the mass centre (Nm)	For bulk material "A" (0.656308; −47.820212; −1151.888135)
		For bulk material "B" (2.639079; −191.280431; −4607.557001)
5	The total reaction of supports (N)	For bulk material "A" (−3057.227214; 0.157531; −1.370206)
		For bulk material "B" (−12228.908843; 0.630143; -5.480826)
6	The absolute value of the reaction (N)	For bulk material "A" 3057.227525
		For bulk material "B" 12228.910087
7	The absolute value of the moment (Nm)	For bulk material "A" 1152.880514
		For bulk material "B" 4611.526502

were made for the average statistical and widely used materials in the construction. The finite element method gave an approximate, nocturnal solution for the physically and geometrically linear rods of feed elements of the screen, greatly simplifying the final system of equations, determining and showing the reliability of the chosen design.

The main test of any calculation results is physical experiment. Considering that the above simulation and calculations represent only feed elements simulation of the real design and how accurate the model and mathematical apparatus implementing this model depend on the experimental verification results. Therefore, on the results of the finite element analysis, a natural, industrially applicable platform with feed elements was installed (Fig. 13). The vibroengine is mounted on the platform with the possibility of changing the amplitude and frequency oscillation, tested for durability and reliability in actual production conditions. A comparison was made between the calculated and actual screen weediness, the dependence of sieve weediness on feed elements number, the sieve weediness dependence on the number of FEs rows, the actual and calculated screen weediness was analyzed, by varying the sieve oscillation frequency.

In the experiment, material (2) moving along the sieve (1), under the influence of vibrations generated by the vibration engine (5), is divided into upper and lower fractions. The elastic supports (4) provide the mobility of the box with the screen (1)

Fig. 13 Industrial platform with feed elements (1 row), installed in the trough of the vibroscreen on the basis of data analysis of finite element modelling

for generating oscillatory movements. The vibroengine works at a constant speed. Particles of bulk material (2) moving along the screen 1 run into feed elements (7) fixed to the frame (6), which leads to the creation of additional chaotic movements of particles in the upper layers of bulk material (2), either coinciding or differing from the direction of the forced oscillations reported by the engine (5).

This increases the difference in the speed of bulk material particles relative to each other, which activates the process of passing particles of lower fraction to the screen surface through the entire layer of bulk material, which increases the screening process efficiency, as a result, increases the screening capacity while maintaining quality sorting.

To determine the efficiency of the FEs operation, it is necessary to identify the most efficient operating mode without FE. For this purpose, the following parameters varied: the amplitude of the sieve A oscillation (0.75–3.0 mm) and the vibration frequency v (8–50 Hz).

After revealing the most rational operating mode of the vibration screen without FE, it is necessary to determine the most effective operating mode of the screen with FE, for this purpose the following parameters were changed: the number of FEs in a row from 1 to 10; the number of FEs rows from 1 to 5.

In this case, the EE is installed at a distance of 200 mm from the beginning of the screening and at a height of 10 mm from the screen, these parameters are established in preliminary experimental studies.

The experiment results on determining the influence of the FE number on the screen weed are given in Table 5, and the graph of the sieve weed dependence on the number of FEs is shown (Fig. 14) .

Further, tests are carried out with a variation of the FE series (Table 6) and an experimental dependence is illustrated (Fig. 15).

On the graph (Fig. 15) the following processes are shown: the lowest index of contamination is reached with the number of rows—from 2 to 10, thus, with the same indices of the minimum level of contamination, a decision is made for further experiments of 2 rows of FE with 5 rods in each row.

Table 5 Test results with different number of FE

FEs number	Weight (g)		Screening time (min)	Weediness of the upper-class Z_B (%)
	Above the sieve	Under the sieve		
1	15,112	6,378	1.40	8.31
2	15,000	6,340	1.38	7.01
3	14,650	6,150	1.40	6.62
4	14,770	6,280	1.36	4.55
5	14,830	6,240	1.36	4.31
6	14,770	6,280	1.36	4.91
7	14,650	6,150	1.34	5.92
8	14,760	6,310	1.34	7.71
9	14,760	6,310	1.33	8.59
10	15,100	6,230	1.32	10.56

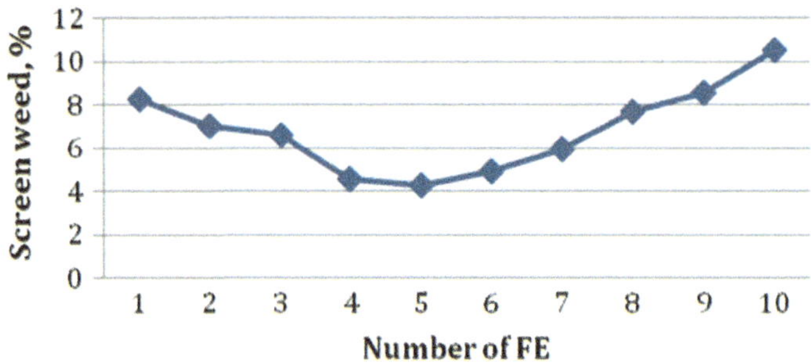

Fig. 14 Graph of the sieve weediness dependence on the number of feed elements

Table 6 Test results with different number of FEs rows

FEs number	Weight (g)		Screening time (min)	Weediness of the upper-class Z_B (%)
	Above the sieve	Under the sieve		
1	14,830	6,240	1.36	4.31
2	14,530	6.000	1.36	3.5
3	14,550	6,001	1.36	3.51
4	14,530	6,002	1.36	3.52
5	14,489	6,590	1.36	3.79
6	14,552	6,590	1.36	4.52
7	14,466	6,590	1.36	4.78
8	14,588	6,587	1.37	5.59
9	14,003	6,004	1.35	6.98
10	14,554	6,001	1.35	7.99

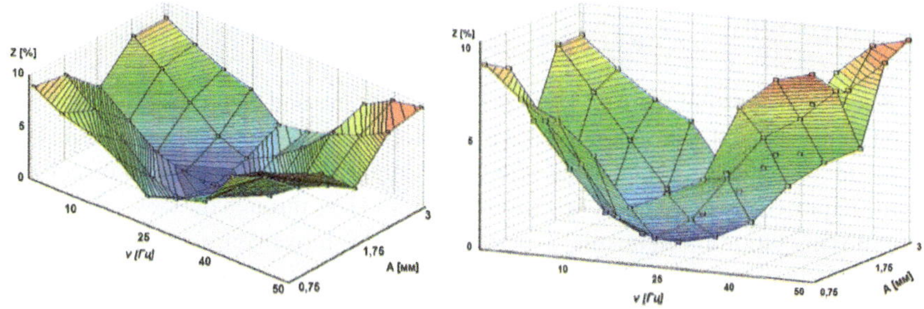

Fig. 15 Graph of the screening weed dependence against the frequency of sieve oscillations, the amplitude of sieve vibrations and the number of FEs, under different inclination angles of the coordinate axes

The evaluation of the strength and rigidity characteristics of the structure (stresses, deformations, displacement of various points in the structure) by finite elemental analysis is fully confirmed in industrial design tests.

To represent the experiment results, multidimensional graph is shown (Fig. 15); the same 3D image of the screening weed, for greater clarity, is shown at different angles of inclination.

The results of experimental studies show a reduction in the level of the upper fraction contamination up to 3.5%, which confirms an increase in the productivity of the screening process, with the same energy consumption, which leads to the conclusion that it is decreasing.

Thus, according to the experimental results, the most effective mode of the vibroscreening process with a 3.5% weed was detected, with an amplitude of sieve vibrations $A = 2$ mm, a sieve vibrational frequency $v = 25$ Hz, a number of FEs in the FEe series of 5 and the number of FEs rows equal to $N = 5$, the time of material passage along the screen was accelerated to $t = 1.36$ min. Consequently, with a normative screen weed of 5%, the productivity of the screening process increased by 2.1%, and the process time decreased by 4.4%.

Experimental studies confirmed the theoretical premise that the additional material stimulation on the screen, when compared with the standard screening on a flat screen, provides an increase in the productivity of the screening process by 2.1%, a decrease in the material passage time through the screen by 4.4%, while maintaining normative weed of the material is 5%.

The practical significance of the work is confirmed by the using of prototypes of the vibrating screen with additional feed elements in LLP "CS RNP" in Ust-Kamenogorsk in 2017.

4 Conclusion

1. Based on the results of advanced experimental studies and theoretical calculations an improved cell screening process model for bulk materials was proposed. The model effectively simulates the screening kinetics, as well as takes into account the nature of the particles penetration of various sizes through the sieve, diffusion and segregation mechanisms of passage particles movement in a forced vibro-fluidized layer.
2. The mathematical and physical model of the experimental installation of the new vibroscreen with additional feed elements was created. In accordance with the developed mathematical model of the cell screening the numerical studies to identify the effect of process parameters on the screening kinetics and the evolution of the state of passage particles in the granular layer were performed. The introduction of additional feed elements to the flow of screened material leads to an increase in the transmission intensity of lower class of bulk material to the sieve. The probability of grains falling to the pan appears earlier in time by 6.66%— in case of static position of the feed elements, by 13.33%—in case of dynamic position of the feed elements.
3. Theoretical and experimental comparison of different screening methods has shown that the use of additional feed elements has a significant effect on the screening process kinetics. It increases productivity and the effectiveness of screening process by reducing the screening time (acceleration of the first screening phase). Thus, new type of vibroscreen can be used in the industry.
4. The simulation model with the finite element method, in contrast to full-scale manufacturing, allows determining the «weak» places in the design at the design stage and approaching the task of optimal parameters selection. The finite element analysis of the feed elements platform is the best and accurate method of researching and predicting the operability of the structure under given operating conditions, allowing selecting the parameters of the future design reasonably prior to its industrial manufacture.
5. The application of the mathematical apparatus (FEM) simplifies the construction of an object model consisting of a finite elements set. FEM allows obtaining a solution in the form of stress and strain fields in practically any section of the element. These advantages of the method have not yet been used in the design of vibroscreen elements. Their implementation can reduce the metal equipment consumption, increase the reliability of its operation and reduce self-cost and, ultimately, improve the quality of the sorted material.
6. Computer simulation technology with the help of FEM allows reliable determination of the real operational characteristics of products, helps customers to ensure that their products comply with the necessary requirements and standards.
7. The results of experimental studies of FEs, produced according to the results of the FEM analysis, showed a decrease in the level of sieve weed up to 3.5%, which allows increasing the productivity of the screening process and also reducing the energy consumption.

8. Experimental studies confirmed the theoretical premise that the additional material stimulation on the screen, when compared with the standard screening on a flat screen, provides an increase in the productivity of the screening process by 2.1%, a decrease in the material passage time through the screen by 4.4%, while maintaining normative weed of the material is 5%.

References

1. Kim, A., Doudkin, M., Vavilov, A., Guryanov, G.: New vibroscreen with additional feed elements. Arch. Civil Mech. Eng. **17**(4), 786–794 (2017). https://doi.org/10.1016/j.acme.2017.02.009
2. Stryczek J., Banaś M., Krawczyk J., Marciniak L., Stryczek P.: The fluid power elements and systems made of plastics. Procedia Eng. **176**, 600–609 (2017) (Elsevier Ltd.). www.elsevier.com/locate/procedia
3. Doudkin, M., Pichugin, S., Fadeev, S.: Contact force calculation of the machine operational point. Life Sci. J. **10**(39), 246–250. http://www.lifesciencesite.com (2013)
4. Doudkin, M., Vavilov, A., Pichugin, S., Fadeev, S.: Calculation of the interaction of working body of road machine with the surface. Life Sci. J. **133**, 832–837. http://www.lifesciencesite.com (2013)
5. Giel, R., Młyńczak, M., Plewa, M.: Evaluation method of the waste processing system operation. In: Risk, Reliability and Safety: Innovating Theory and Practice—Proceedings of the 26th European Safety and Reliability Conference, ESREL 2016, Glasgow, Scotland, 25–29 Sept, pp. 912–917
6. Fedotov, A.I., Młyńczak, M.: Simulation and experimental analysis of quality control of vehicle brake systems using flat plate tester. In: Dependability Engineering and Complex Systems: Proceedings of the Eleventh International Conference on Dependability and Complex Systems DepCoS-RELCOMEX, Advances in Intelligent Systems and Computing, June 27–July 1, 2016, Brunów, Poland, pp. 135–146
7. Bojko, A., Fedotov, A. I., Khalezov, W. P., Młyńczak, M.: Analysis of brake testing methods in vehicle safety. In: Nowakowski, T. et al. (eds.) Safety and Reliability: Methodology and Applications. Proceedings of the European Safety and Reliability Conference, ESREL 2014, Wroclaw, Poland, 14–18 Sept 2014, Leiden: CRC Press/Balkema, cop, 933–937 (2015)
8. Surashev, N., Dudkin, M., Yelemes, D., Kalieva, A.: The planetary vibroexciter with elliptic inner race. Adv. Mater. Res. **694–697**, 229–232. Trans Publications, Switzerland (2013). https://doi.org/10.4028/www.scientific.net/AMR.694-697/229
9. Doudkin, M.V., Fadeyev, S.N., Pichugin, S.Y.: Studying the machines for road maintenance. Life Sci. J. **10**(12s), 134–138. http://www.lifesciencesite.com (ISSN:1097-8135) (2013)
10. Doudkin, M.V., Pichugin, S.Y.U., Fadeyev, S.N.: The analysis of road machine working elements parameters. World Appl. Sci. J. **23**(2): 151–158 (2013). IDOSI Publications. https://doi.org/10.5829/idosiwasj.2013.23.02.13061 (ISSN/E-ISSN: 1818-4952/1991-6426)
11. Doudkin, M., Kim, A., Kim, V.: Application of finite elements method for modeling and analysis of feed elements stresses of the vibroscreen. In: Proceedings of the 14th International Scientific Conference: Computer Aided Engineering,8 p. Politechnika Wrocławska (2018)
12. Detyna, J.: Maximum entropy as a theoretical criterion of statistical description of the granular matter separation. Publisher House Oficyna Wydawnicza Politechniki Wroclawskiej, Wroclaw (2007)
13. Molerus, O.: Stochastisches Modell der Gleichgewichtsichtung. Chem. Ing. Tec. **39**(13), 792–796 (1967)
14. Ogurtsov, V.: Stochastic model of screening particles distribution in the layer of particulate material during vibration screening. Build. Mater. **11**, 38–39 (2007)

15. Kim-Vainberger, A., Doudkin, M., Vavilov, A., Guriyanov, G.: Inventors; Assignee. Vibrating screen. Innovative Patent of the Republic of Kazakhstan No. 31145. 17 April 2015
16. Detyna, J., Bieniek, J.: Methods of statistical modeling in the process of sieve separation of heterogeneous particles. Appl. Math. Model. **32**, 992–1002 (2008)
17. Detyna, J.: Stochastic models of particle distribution in separation processes. Arch. Civil Mech. Eng. **10**, 15–26 (2010)
18. Zhuravlev, A.: Condition of nonmetallic building materials industry and its development prospects. Building Materials **11**, 4–6 (2007)
19. Doudkin, M., Kim, A., Kim, V.; Application of FEM method for modeling and strength analysis of FEED elements of vibroscreen. In: Proceedings of the 14th International Scientific Conference on Computer Aided Engineering, June 2018. Series: Lecture Notes in Mechanical Engineering, 892 p. Wroclaw, Poland (2019)
20. Weisberg, L.: Design and calculation of vibrating screens. Publisher House Nedra, Moscow (1986)
21. Doudkin, M., Kim, A., Kim, V., Mlynczak, M., Kustarev, G.: Computer modeling application for analysis of stress-strain state of vibroscreen feed elements by finite elements method. In: Proceedings of Mathematical Modeling of Technological Processes International Conference, CITech 2018, Ust-Kamenogorsk, Kazakhstan. 25–28 Sept 2018, pp. 82–96. https://doi.org/10.1007/978-3-030-12203-4_9

The manufacturer's authorised representative in the EU is Springer Nature Customer Service Centre GmbH, Europaplatz 3, 69115 Heidelberg, Germany. If you have any concerns regarding our products, please contact ProductSafety@springernature.com

Printed and bound by CPI Group (UK) Ltd, Croydon, CR0 4YY

23/07/2026

02175027-0002